INSTITUTIONAL BARRIERS TO SUSTAINABLE TRANSPORT

Transport and Mobility Series

Series Editors: Richard Knowles, University of Salford, UK and Markus Hesse, Université du Luxembourg and on behalf of the Royal Geographical Society (with the Institute of British Geographers) Transport Geography Research Group (TGRG).

The inception of this series marks a major resurgence of geographical research into transport and mobility. Reflecting the dynamic relationships between socio-spatial behaviour and change, it acts as a forum for cutting-edge research into transport and mobility, and for innovative and decisive debates on the formulation and repercussions of transport policy making.

Also in the series

Daily Spatial Mobilities
Physical and Virtual
Aharon Kellerman
ISBN 978 1 4094 2362 1

Territorial Implications of High Speed Rail
A Spanish Perspective
Edited by José M. de Ureña
ISBN 978 1 4094 3052 0

Sustainable Transport, Mobility Management and Travel Plans
Marcus Enoch
ISBN 978 0 7546 7939 4

Transition towards Sustainable Mobility
The Role of Instruments, Individuals and Institutions
Edited by Harry Geerlings, Yoram Shiftan and Dominic Stead
ISBN 978 1 4094 2469 7

Integrating Seaports and Trade Corridors
Edited by Peter Hall, Robert J. McCalla, Claude Comtois and Brian Slack
ISBN 978 1 4094 0400 2

For further information about this series, please visit www.ashgate.com

Institutional Barriers to Sustainable Transport

CAREY CURTIS
Curtin University, Australia

NICHOLAS LOW
University of Melbourne, Australia

ASHGATE

Published by
Ashgate Publishing Limited
Wey Court East
Union Road
Farnham
Surrey, GU9 7PT
England

Ashgate Publishing Company
Suite 420
101 Cherry Street
Burlington
VT 05401-4405
USA

www.ashgate.com

British Library Cataloguing in Publication Data
Curtis, Carey.
Institutional barriers to sustainable transport. –
(Transport and mobility series)
1. Urban transportation – Environmental aspects. 2. Urban transportation – Social aspects. 3. Urban transportation policy – Australia – Case studies.
I. Title II. Series III. Curtis, Carey.
388.4–dc23

Library of Congress Cataloging-in-Publication Data
Curtis, Carey.
Institutional barriers to sustainable transport / by Carey Curtis and Nicholas Low.
p. cm.
Includes bibliographical references and index.
ISBN 978-0-7546-7692-8 (hardback) — ISBN 978-0-7546-9390-1 (ebook)
1. Urban transportation—Environmental aspects. 2. Urban transportation policy. 3. Urban transportation—Social aspects. 4. Sustainable development. 5. Urban transportation policy—Australia.
I. Carey, Curtis. II. Title.
HE305.L677 2012
388.4'042–dc23

2012012829

ISBN 9780754676928 (hbk)
ISBN 9780754693901 (ebk – PDF)
ISBN 9781409484103 (ebk – ePUB)

MIX
Paper from responsible sources
FSC® C018575
www.fsc.org

Printed and bound in Great Britain by the MPG Books Group, UK.

'Cahill Expressway exit ramp, Sydney – a subtle message to transport policy makers?' Photo: Carey Curtis

If things need to change look to yourself first

Contents

List of Figures

List of Tables

Notes on Authors

Carey Curtis is Professor in City Planning and Transport at Curtin University, Australia. She is a Partner of the Australasian Centre for the Governance and Management of Urban Transport and Chairs the International Editorial Board of the journal *Urban Policy and Research*.

Nicholas Low is Professor in environmental planning at the University of Melbourne, Australia. He is Associate Director of the Australasian Centre for the Governance and Management of Urban Transport (GAMUT). He is a member of the International Editorial Board of the journal *Urban Policy and Research*.

Preface

Much research has answered the question 'what can make economic activity on this planet sustainable?' Yet we see ever growing depletion of the biosphere. It is time to ask what is *stopping* sustainability? This book is the culmination of a research project seeking an answer to that question in the field of transport and urban planning. The purpose of the book is to explain the importance of local and national institutional change in a world seeking to tackle global environmental problems such as climate change. Through an examination of transport planning in Australia, the book challenges the conventional wisdom by showing, through original research, how 'car dependence' is as much an institutional as a technical phenomenon.

To the engineers and consultants who have borne the major responsibility for transport planning since the beginning of the last century we say this. We salute your work and the immense professional competence you have displayed in carrying it out. The vision of road engineers was the vision of many in twentieth century society. Railway engineers worked hard to serve the demands of politicians to reduce dependence on the budget. The theory of path dependence specifically avoids any idea that the agents who diligently carry out the work of governments are to blame for following a path which all their intelligence tells them is desired by society. All we ask is for you now to look closely at where that path is leading and apply your undoubted problem solving skills to shaping a new path that will better serve to give cities and the planet a future.

Case studies in three metropolitan cities show how Australian transport policy has become institutionally fixated on a policy path dominated by private, road based transport. Exploration of key documents over some fifty years, and interviews with key actors, demonstrate the existence of storylines supporting infrastructure for cars and militating against investment in public transport and other forms of movement. Our research shows how policy systems become encrusted around investment to accommodate private cars, erecting an impenetrable barrier against more sustainable mobility and accessibility solutions.

The book represents a new approach to understanding transport policy that brings political-institutional analysis to what has traditionally been the domain of engineering and technology. The way its empirical content is connected to this theory and the issue of sustainability makes its findings applicable to most cities of the developed world, and to fields beyond transport planning.

Carey Curtis and Nicholas Low

Acknowledgements

The writing of this book was a collaborative effort of the two authors which merged insights and in which we mutually learned from and supported each other. However, we each have personal things we want to acknowledge which really can't be said jointly. Moreover there were different helpers in Perth and Melbourne, and at different times, during the research. So this section of the book is the only one in which we make separate contributions.

Carey Curtis

Among the significant things I learnt, in 2010, was the importance of acknowledgement. It was suggested that the thing you most want for yourself is the thing you are least able to give to others. So here I want to rectify any possible failing on my part by acknowledging just how much support and love my partner David gives to me. To my son Janni, who from a very early age picked up the planning challenge to turn our cities towards a more sustainable path, I enjoy our interesting policy debates! I also want to acknowledge David's son, Simon Nicholson (1994 – 2010). In Simon's case we worked hard to give him the acknowledgement he so needed. Perhaps I saw in him what I have always yearned, but for Simon our acknowledgement was not enough – he longed for someone closer to him to meet his needs. May you rest in peace Simon for you are sadly missed by all those close to you. We will never know what gifts you had yet to bring.

Aside from my personal acknowledgements, I must also thank my fellow author. Nick's persistence in seeking a research grant saw our work develop. An Australian Research Council Grant (DP0450709) enabled the study of the three Australian cities. We have made equal contributions in this book and the research that it draws from. I have drawn on Nick's strength as a theorist and political scientist and in return my contribution has been to the empirical research – systematizing the methodology to enable data collection across three cities (each with different institutional structures and processes) and analyzing a large body of data and synthesizing this across the three cities.

I thank the research assistants who supported my contribution to this research for different periods during the three year project: Carlindi Holling, Melanie Montgomery, Diane Daze, Anthony Kent. Without their diligence and persistence in gathering data the possibility of us bringing this book to fruition would have been lost. Oscar Thomson assisted with figures and referencing. I also acknowledge the wise counsel I have received from fellow transport and planning practitioners in Western Australia. They are the people I can turn to for local and historical knowledge about the way in which transport planning and policy has taken shape. In this respect I thank Ian Ker, Stuart Hicks and Ian Alexander; each has been active in transport and city planning for a large part of the study period. I thank Glen Searle for his guidance with the Sydney case. I thank my colleagues at Curtin University for the support provided by protecting my time so that I could work on this book.

Nicholas Low

First, on a personal note, I want to say that Elizabeth, my wife, has always been tolerant of the inevitable strains that occur when I'm writing a book. When writing, I'm often conscious of Liz metaphorically looking over my shoulder. With a Melbourne University degree (English and Geography majors) and a career behind her as a secondary teacher, she is a good critic. Now she is a ceramic artist of distinction (Liz Low) and I hope I have been able to return the support for her work that she has given me in mine.

Carey has been a good friend and colleague, and I depend on her not least for the methodological rigour she brings to empirical research. But of course there is much more than that to thank her for. She brings to our work both theoretical insights and high level practical experience of urban planning as a member of the Western Australian Planning Commission. For some time now Carey has been one of Australia's leading planning scholars. We both come from a planning background and support each other in the conviction that transport planning is too important to city futures to be left entirely to transport planners.

There are two scholars who I also particularly want to thank for their contribution to the research. First, Professor Brendan Gleeson was a principal investigator in the ARC Discovery Project (DP0450709) which was the core of the research. Because of his own workload he could not contribute to the book, but I warmly acknowledge his part in the research. The second is Professor Bill Russell whose work on the federal role in transport planning I greatly value and who is the co-author of Chapter 4. This book could not have been written without their contribution. My thanks also go to those who enjoy and support research at the Faculty of Architecture, Building and Planning of the University of Melbourne: both scholars and professional staff.

Institutional benefactors I want to thank are, first (in time), The University of Melbourne which awarded me small grants in 2001 and 2002 that helped provide the foundation for the successful bid to the Australian Research Council. I also want to thank the Volvo Research and Educational Foundations which funded the GAMUT Research Centre in Sustainable Urban Transport from 2006 to the present, of which I was Director up to the end of 2010. GAMUT and its partners (Carey is one) helped me to view future urban transport in a broader and deeper perspective.

Most important to the research have been in approximately chronological order, Dr Emma Rush, who assisted the empirical research in the early years; Ms Abbe Villis who worked on the interviews with key policy shapers in Melbourne, and Ms Rachel Astle who took over from Emma in the later years of the project. They are meticulous researchers and in working with them I gained many insights which truly enriched the research. I also want to thank and acknowledge the splendid drafting of diagrams and figures by Ms Chandra Jayasuriya, cartographer at the University of Melbourne.

Finally we both thank Valerie Rose at Ashgate for her patience, support and understanding when, with the events of 2010, the book took longer to complete than we had planned. Valerie provided a much needed sense of perspective at a critical time.

Copyright permissions:

Chapter 1
Introduction

The twentieth-century vision of urban transport contained five crucial elements. First, the vision itself was of unimpeded, individual, motorized mobility. Along with this promise of individual mobility, went speed. Vehicles should at all times be able to travel at speeds consistent both with their design capacities and a reasonable level of safety for the drivers and passengers. Second, however, the attempt to realize this promise for all who could drive provoked traffic congestion, which then became the dominant transport problem of the twentieth century. Time lost moving at less than the optimum speed was time, and therefore money, wasted: the cost of congestion. Third, the solution to traffic congestion was to build more and better roads to accommodate private vehicles moving at speed. The higher the speed of the vehicles, the greater the separation between them necessary for the safety of people in cars, and so the more road space was required. Fourth, as realization of the promise became a historical fact, a simple technology was developed to project the growth of 'demand' for individual motorized mobility, and predict the road space needed to meet it. The fifth and final element was the implicit assumption that the planetary environment would for ever provide the necessary resources to fuel the vision and absorb its wastes.

This vision is powerful because it taps the human urge for freedom. 'Of all the specific liberties', wrote the philosopher Hannah Arendt, 'which may come into our minds when we hear the word "freedom", freedom of movement is historically the oldest and also the most elementary. Being able to depart for where we will is the prototypical gesture of being free' (Arendt, 1955: 9). The first half of the twentieth century was full of the ideology of human emancipation by the private car (see Davison, 2004: 112–116). The vision is strongly compatible with liberal capitalism: freedom of consumer choice, the free market and, above all, the free movement of labour. The transport vision grew within the shell of free market economics, which in turn emerged from the powers that shaped the eighteenth century philosophy of the Enlightenment. So perhaps it is fair to talk of the 'traffic enlightenment'.

The twentieth-century vision has nowhere been achieved, not even in that urban icon of the twentieth century, Los Angeles. Indeed the *Urban Mobility Report*, 2010, of the Texas Transportation Institute which tracks congestion in American cities reports Los Angeles as the most congested of the 15 largest US conurbations with 515 million hours travel delay in 2009 (Texas Transportation Institute, 2010). But in England the transformation of the urban environment necessary for its achievement was considered in the 'Buchanan Report', *Traffic in Towns*. In a portion of central London (east of Euston Road and north of Tottenham Court

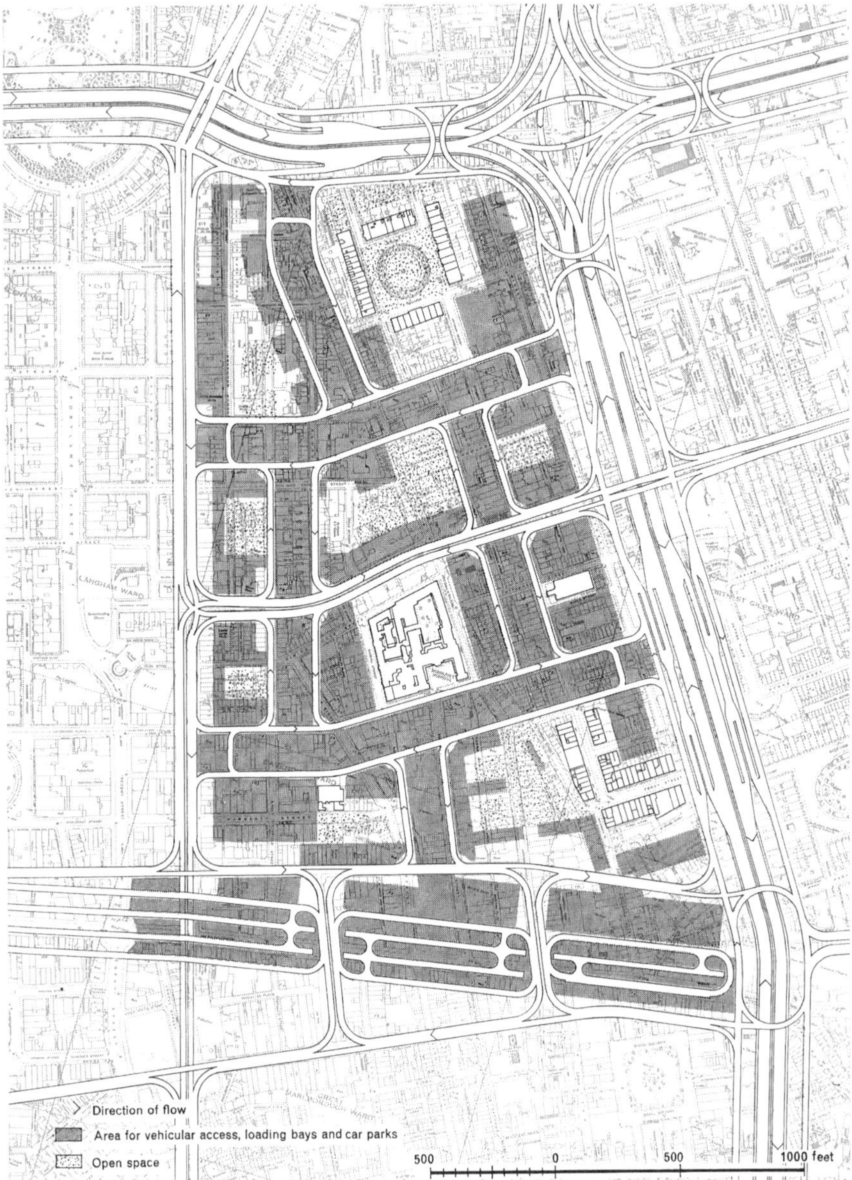

Figure 1.1 Buchanan's design for partial redevelopment

Source: *Traffic in Towns, A Study of the long term problems of traffic in urban areas*, Report of the Working Group (led by Colin Buchanan) HMSO 1963 Figure 192, Page 147 'The design for partial redevelopment – plan at ground level showing the primary, district and local distributor road system together with the parking and service areas.'

Road), the Buchanan team calculated the consequences 'if every person should seek to go to work by car, every shopper to use a car, and the residents to have all they desired in the way of cars and parking spaces' (Working Group, 1963: 130). Their conclusion: 'when we considered the consequences for the primary [road] network of a vast continuous spread of areas similar to our study area, for this is what the middle of London really comprises, we realized that the network would become *impossibly* large and complicated' (ibid.). Nevertheless the Buchanan team decided to try out a compromise version: not what might be needed within the area itself, but 'what could be practically contrived in the way of a network to bring traffic to and from the area' (ibid.). The radical transformation required for even this compromise is shown in Figure 1.1. It was a shocking wholesale demolition of the urban fabric, and its reconstruction around a pattern of motorways and 'urban rooms' segregated from the main flows of road traffic.

At this high point of the 'traffic enlightenment' in the 1960s, when popular belief in the virtue of the private car was strongest, the modernist architecture of reconstruction was still fashionable, the level of oil field discovery was peaking, and the 'environment' meant merely the local physical space, the Buchanan Report was a tipping point of sorts, or perhaps a Janus gateway between the past and the future, facing two ways, at once recognizing the impossible cost of the vision yet still hankering after its partial realization.

Throughout the latter half of the twentieth century in the developed world, there were many complaints and protests launched against implementation of the 'traffic enlightenment' vision of motorized individual mobility. But they gained little traction. Commenting on the Buchanan Report, Peter Self, then chairman of the Town and Country Planning Association, sensibly observed, 'Should we not utilize the best means of transport for each particular purpose? For city centres, this would imply heavy use and almost complete reliance upon public transport, which is particularly well suited to this task. Conversely, the best prospects for motorization surely lie in the conception of a regional complex of new and expanded towns, each relatively small' (Self, 1963: 1228). But by the 1960s an institutional carapace of organizations, policies, beliefs and practices had, throughout the developed world, been built around the vision, protecting it from challenge from without.

It was never enough to show that implementation of the vision had serious negative side effects or externalities. It was never enough to show that the vision was not equally applauded by everybody, that some sections of the population were losers from its pursuit. It was not enough to point out the vision's large opportunity costs. It was not enough to demonstrate that pursuit of the vision in urban areas was destructive of the quality of the urban environment. It was not enough to show that building or improving roads only added to traffic on the road system. The policy path was sufficiently entrenched to brush off such challenges. Despite more subtle arguments to the contrary, the conventional wisdom continues to assert that fighting congestion is in the public interest, that the public are voting with their wallets to buy and use cars, and that the rest are vested interests with axes to grind. If the result is traffic congestion, the common sense solution is to build more road space to free up bottlenecks. It is politically 'courageous' to oppose such popular trends.

What has finally exposed the glorious vision as a lying mirage are two discoveries of the last century whose consequences have yet to be fully confronted by the state apparatuses of our times. These are the catastrophe of climate change induced by combustion of fossil fuel (Weart, 2003, provides a fine and readable account of the discovery of human induced climate change), and the peak and subsequent decline of the oil supply (Deffeyes, 2002; Campbell, 2005).[1] Whatever the individual pleasures – and they are many – of the use of private motor vehicles, a vision in which their use is the dominant mode of urban transport cannot be defended when only a small rich minority will be able to afford it, and when its pursuit causes climate change. The world today stands on the threshold of a mobility catastrophe. The peak of oil production and climate change immediately confront the world, and governments around the world have yet to offer a robust and truly effective solution. It is not enough to argue that vehicles will have to change. Rather *urban transport*, and possibly our habits of mobility, will have to change. The forces preventing that change are now shown to be deeply irrational.

Many different kinds of values, some of them contradictory, are usually rolled up in the term 'sustainable' (Low, 2003). They will not be debated here. Suffice it to say that a transport vision that does not recognize and is not adapted to the realities of climate change and peak oil is not sustainable. The purpose of this book is to report an investigation of the institutional barriers preventing the development of a new vision compatible with these realities and in those terms 'sustainable'.

Research Approach

This book draws on a major research project which set out to examine one particular policy domain, transport, in order to examine how discursive and institutional structures within government influence the achievement of sustainability targets. Our research questioned how far and fast government agencies move policy towards environmental sustainability. We argued that this depends not only on plans and programmes espousing the goal of sustainability, but also on the existence of barriers to implementation, and opportunities for innovation, resulting from existing discursive and institutional structures. The central hypothesis is that 'discourse networks' play a key role in the paradigm shift entailed by 'sustainable development'.

The 'discourse network' is a concept used to explain the interconnectedness of ideas, decision-makers and their mental models of reality: what is 'important',

1 There is a debate about whether the peak of oil production has passed or is still to come. But the facts are mired in the politics of oil. There is really no *scientific* debate worth mentioning about the facts of climate change.

what 'the problem' is, and how to go about solving it. For example, existing transport policy (with an emphasis on road building and private vehicle traffic) remains essentially unchanged so long as key actors support it. These actors do not have a unique idea of why it is necessary to support it but rather express a variety of ideas, or 'storylines' (following Hajer, 1995) that provide reasons for current policy solutions and may also suggest directions for change. : Hajer proposes that influence flows two ways: institutions manifest their existence through the intentional acts of their members but they also form and govern those intentions, partly by influencing the conceptions on which their members act (Hajer, 1995: 42–72). These conceptions are shaped by particular belief systems or 'storylines' (Davies and Harré, 1990). Storylines are metaphors, analogies, historical references and clichés, which hold discourse together, but they are also 'prime vehicles of change' (Hajer 1995: 63). He means that the development of new policy requires the incursion of new storylines.

As the research proceeded it became clear that what we were uncovering was a condition of institutional stasis, a failure to respond adaptively to the key challenges of the twenty-first century. To understand that condition we needed to set Hajer's conceptual insights within the broader conceptual framework of institutional path dependence mapped out by theorists such as North (1990), Pierson (2000 a and b, 2004) and Torfing (2001).

Our empirical focus is the transport policy systems of three Australian cities: Melbourne, Sydney and Perth. We show how, over a period of 50–60 years, each has been extremely resistant to adoption of the paradigm of sustainability, except in rhetoric.

Our research design comprised four parts. First, using content analysis, we analysed the policy discourse of State and federal government policy documents in the period from the making of the first major land use and transport plans and policies in the late 1950s to around the year 2000. We conducted this for road infrastructure and for public transport in each city. In addition we analysed documents put out by key non-government actors (pressure groups): both pro-road groups and pro-public transport. Second, we turned to case studies of infrastructure projects in order to deepen the investigation. We focused on three major road building projects (one in each city). We examined the environmental impact statements and other planning reports, mapping arguments and investigating their logical coherence. Third, we analysed the institutions, looking at secular change in organization charts and the rules and procedures of agencies. Finally, we conducted a series of in-depth interviews with key informants, or policy shapers, in each city. These included those who had key influence over road infrastructure and public transport decisions. They included past and present Ministers of transport and land use planning, heads of State Government roads, public transport and planning agencies, persons responsible for case study projects and leading figures in pressure groups.

Structure of the Book

A 'sustainable' transport system means one that meets a variety of goals other than simple mobility. Some are rather intangible such as improved quality of urban space. Others, such as distributional equity in transport, are rarely measured, though increasingly important. In Chapter 2 we first briefly review the challenges of sustainable transport. We share the view put forward by Pacala and Socolow on the climate problem (2004: 968) that a 'portfolio of technologies' currently exists that can meet the world's energy needs while stabilizing greenhouse emissions ('reduced reliance on cars' being one option). These authors specifically address the question of climate change, but we argue that the same holds for other major challenges of sustainability such as oil depletion and risk to health. We review the contents of a green transport portfolio and show that it is feasible. What then is blocking progress towards this end? As Banister (2005: 93) points out, 'the role of institutions is crucial to the effective implementation of challenging options on sustainable development'; but time and again it appears that institutions block the way.

If politics in a democracy is an ongoing game, 'institutions' are the game's rule book – the organizations set up to devise solutions and implement decisions, the norms, values, beliefs, assumptions and even language which imbue the process of decision-making. A key to understanding institutional behaviour, especially 'policy lock-in' (mentioned in the Stern Review p. 397) is the theory of 'path dependence'. Chapter 3 explains in practical terms what this theory means for transport planning. The interpretation of path dependence explored in this chapter considers three related elements. The first is the path created for cities by the weight of physical infrastructure: the form an urban transport system takes over a period of time shapes the future of a city (Arthur, 1988; Woodlief, 1998). The second is the organizational power of the agencies making transport policy: policy is determined by agency power (Vigar, 2002; Dudley and Richardson, 2002). The third is the discourse that defines the transport problem for policy makers: policy is shaped and supported by arguments, perspectives and storylines (Low, 2005). The empirical focus of this book is the second and third elements: institutional path dependence.

The continuity of public policy from one political administration to the next is something most people take for granted and is accounted for to a considerable degree by path dependence. Root and branch change of policy with each change of government would be chaotic if it were not also impossible. So path dependence need not be seen as necessarily a policy failure. Rather, path dependence turns into policy failure when the path of policy development becomes inconsistent with the long term public interest. This failure is symptomatic of the insufficient capacity within a polity for long term policy review and redirection. The case study of this book is Australia. Current transport policy in Australia, like that in much of the developed world, is negatively path dependent and lacks the capacity for review and redirection.

Australian urban development over the last 50 years, characterized by low density peripheral growth loosely governed by planning laws, has had much in common with the development of the cities of North America and Europe. In the first part of Chapter 4 we briefly set the Australian scene in international context, drawing attention to the particularity of Australian urban geography and political constitution. In the second part we describe the trajectory of development of Australian transport planning with particular emphasis on the important role of the federal tier of government and how federal funding and policy shaped infrastructure at State level. Australia is a federal state and the funding for transport is shared between the federal (Commonwealth) and State governments. We draw attention to the key difference between public transport expenditure which is mostly operational, paid through State budgets, and road expenditure which is mostly capital expenditure, with a substantial portion provided by the federal government. In the final part of the chapter, critical periods ('critical junctures') are identified in the history of Australian cities and their governance that became turning points in urban transport planning.

City form – locational patterns of activity and buildings – is shaped by large scale infrastructure (Arthur, 1988) and this shaping can be termed 'technical' path dependence. For instance in Melbourne the sectoral pattern of development shaped by the railways continued to influence all journeys to work long after rail travel had been replaced by car travel as the dominant mode. However the physical reality of past investment in infrastructure should not be regarded as the only factor determining current transport policy. In the dispersed cities, characteristic of Australia and North America, altering (via planning policy) the urban form and population densities sufficiently to challenge over-reliance on the car is particularly challenging to implement and routinely opposed by community interests.[2] Alternative strategies focused on integrated transport networks are available for dispersed areas (see HiTrans, 2005; Mees, 2000, 2009). Moreover, the substantial differences in policy in the three Australian cities of our case study, even though they share similar histories of road infrastructure investment, suggest the existence of other forms of path dependence shaping transport policy.

Chapters 5 and 6 explore the impact that the historical development of transport planning organizations has had on transport and infrastructure policy. In these chapters we examine the rule-bound social structures that enable collective action, the separate bureaucracies of roads and public transport. In Chapter 5 the development of the road planning agencies is analysed. The approach taken first describes the structural changes that occurred from the early days up to about the year 2000. In Sydney transport planning remains fragmented with roads and public transport agencies operating in separate compartments. In Melbourne the story is one of partial integration – varying at different periods. In Perth, by contrast, there

2 We prefer the term *car reliance* to car dependence because some people actually have to *depend* on the car: the disabled, the infirm and most of us above a certain age. But a great many people who are well able to use other modes of mobility rely on the car either from simple preference or because infrastructure and a competitive service for other modes of travel are unavailable.

has been a strong movement to integrate roads and public transport planning, and also transport with land use planning. We describe the reduction in the number of single purpose institutions (each with its own Minister) as the State governments have tried to integrate the planning of all transport together with land use planning. The examination of legislation, however, suggests that the road planning agency in all three cities retains considerable power, despite organizational change seeking to limit this power.

Following the framework of Chapter 5 the focus turns in Chapter 6 to the history of organizations for public transport planning and delivery. The different structures adopted in the three cities are compared. Sydney has more organizations than Perth, and brings with it the corporatization story and rapidly changing infrastructure policy as transport becomes politicized. Melbourne's history is one of gradual integration of independent public transport agencies up to the 1980s, followed by a radical experiment in privatization that greatly reduced the potential for integrated planning across the three modes: heavy rail, light rail (trams) and buses. In the 1980s Perth saw the final stage of a process whereby many public transport agencies were merged to create a single organization. In all cities, however, public transport agencies had limited input into strategic policy formulation for delivery of the transport network as a whole. In conclusion we assess the historical organizational strength of the public transport agencies compared with those with responsibility for roads.

Turning from organizations to policy and plans, in Chapter 7 we examine the impact the different organizational sectors have had on plans, for both transport and land use. We discuss the integrative turn in public policy in the last ten years. We use the term *integrative* rather than integrated because, at least at first, integration is little more than aspirational. Actual integration of urban and transport policy has far to go in all our case study cities, even though Perth appears to be the leader and Sydney the laggard. Yet it seems that the movement towards integration is gaining momentum and may yet signify a new critical juncture in transport and infrastructure policy.

Chapters 8 and 9 turn to the discursive aspect of path dependence starting with the 'discourse of roads'. Analysis of key policy documents and annual reports produced since the 1950s by roads and urban planning agencies reveals a variety of argumentation supporting expenditure on new and improved roads. The investigation is focused by case studies in the three cities: the development of ring roads in Melbourne, Sydney and Perth. Using the 'storyline' as a conceptual strategy (as in Hajer, 1995) for evaluating the argumentative content of documents, we analyse the discourse surrounding planning to improve the road system. By mapping these storylines over time it is possible to compare their relative strength in the three cities. Storylines supporting discursive propositions change, or are maintained, creating a shared language for the institutional supporters of road investment. Interviews with key informants show how much the storylines elicited from published documents inhabit the minds of policy makers in recent times. In some cases key premises are challenged, but the overall finding is a high level of agreement with propositions reflecting the most influential storylines.

Using a similar strategy to investigate the discourse of public transport, policy documents, annual reports and case studies are analysed in Chapter 9 for the public transport case. In Sydney and Melbourne investment in railways (heavy rail) and trams (light rail) was mostly completed by the early twentieth century. After this the rise of car production propelled a demand, focused by the pro-car lobby, for investment to be shifted to roads. The tramways were abolished in Sydney in the 1950s but retained in Melbourne as a result of the strong leadership of the head of the tramways agency. In Sydney and Melbourne the dominant public transport storyline from the 1960s to the present has been *containing the public transport deficit* – a 'drain on the public purse'. This storyline goes with the *inevitability* story: that the car is the modern method of travel which inevitably replaces public transport in public favour. These stories featured strongly in Perth during the 1960s and 1970s.

Of course the private transport and public transport systems differ in an important way. Whilst the operational expenditure on private transport (owning, driving, maintaining and insuring vehicles) is borne by the individual, the same expenditure on public transport is borne collectively by government. The total aggregate expenditure on private transport (including both operational and capital expenditure) greatly exceeds that on public transport. But economic ideology insists that such individual operational expenditure is the result of a rational historical choice as travellers began to use their cars rather than trains, trams and buses. Yet, over time, this past choice creates a path dependence in which choice is restricted in the present. The downward spiral of public transport provision erodes the choice available such that for certain areas of the city there is no real choice.

Perth and Melbourne have, at different times, undergone shifts in the dominant public transport storyline (reducing the deficit) towards a story that emphasizes *world class public transport*. Yet in all three cities the feeling of inevitability of car-borne transport retains its hold, and public transport agencies still do not see their role as competing effectively with the car.

In Chapter 10, we look beyond the government agencies at some of the surrounding stakeholder groups, comparing the pro-roads groups (Sydney NRMA, Melbourne RACV, Perth RACWA) with pro-public transport and groups opposing road building (Sydney EcoTransit, Melbourne PTUA, Perth TRAC). Again using the storyline as an analytical tool we identify the way in which these groups articulate their message.

The road service agencies in the three cities have consistently complained about what they see as a shortage of roads funding in Australia. They have argued that this deficiency undermines an important national asset: the road system. A storyline frequently used is *road safety*, throughout the decades presented in highly emotional terms. Curiously the rather more realistic storyline of *road risk* to describe the same events has not been employed by supporters of public transport even though the risk to life and limb from road travel is enormously greater than the equivalent risk – in fact all risk – from public transport use.

Most recently, pro-roads agencies in Perth, and to some extent in Melbourne, have joined the public transport bandwagon, arguing that building more and better roads will help to facilitate the movement of public transport vehicles, with freeways seen as particularly beneficial to public transport. In relation to sustainability the pro-roads groups have seen the solutions to the environmental problems resulting from car transport and the unsustainable use of fossil fuels as requiring 'technological fixes' including better emission control equipment, more fuel-efficient vehicles, improved vehicle maintenance, and development of alternative fuels.

The relative strength of the pro-roads group is evident through its membership numbers, income and embeddedness within the institutional structures for transport planning. Though they vary in the three cities (with Melbourne's PTUA being the strongest on conceptual arguments), the pro-public transport and anti-roads groups focus mainly on the environmental storylines, and the social virtues of public transport, but have tended not to tackle the pro-roads storylines directly. Whereas the pro-public transport group in Melbourne can be considered a lobby group outside government, the pro-roads group operates as more of a network overlapping with government and the roads agency.

In Chapter 11 we turn to contemporary mental models. Using interviews with key policy shapers in the three cities, we examine the extent to which the storylines we have elicited from published documents inhabit the minds of policy makers. In some cases key premises are challenged, but the overall finding is a high level of agreement with propositions reflecting the most influential storylines. This is particularly apparent when drawn along the lines of key informant groups (those representing roads, public transport, politicians, town planners and social activists).

Chapter 12 returns to the theoretical level, reviewing the discussion of the previous chapters to consider what sort of a case can be made that transport planning in Australia is path dependent. The dimensions of path dependence are reconsidered and the evidence summarized. The proposition of path dependence is broadly upheld. The technical dependence of cities on car transport has been shaped by a continuous high level of investment in bigger and better roads, which in turn encourages a pattern of land use serviced dominantly by car use. This technical path has in effect created dual cities in Sydney, Melbourne and Perth, with inner and middle regions, broadly, having better (though not optimal) public transport systems and outer regions that are extremely car reliant. This urban form, however, was shaped by a path dependent policy created by a strong unified organizational structure for road planning and construction coupled with a weak and fragmented structure for public transport.

Finally, Chapter 12 considers what might be done to dislocate the policy path that has been so dominant over the last 60–70 years. We outline a strategy and program of action to build on changing public perceptions. An action program is proposed aimed at creating a new vision for urban transport. We envisage a change in the way the community and governments view the future of urban transport. The twenty-first-century vision is one of seamless public transport networks even for dispersed urban areas, a reduced role for the private car for routine city trips, greatly improved facilities for walking and cycling, and integration of land use and transport planning. In this vision, the main issue is seen as the impact of transport on climate change, increased oil prices and eventual reduction in oil supply. The growing public demand is for better public transport and its connection with foot and bicycle paths. The program needs to mobilize public opinion to demand greatly improved management and delivery of public transport in order to create real modal choice for people living in the outer, car reliant urban areas of Australia's cities.

Chapter 2
Sustainable Transport and Institutional Barriers

In common language a 'sustainable' urban transport system is one that meets a variety of goals other than simple mobility. Some are rather intangible such as improved quality of urban space. Others, such as distributional equity in transport, are rarely measured, though increasingly important. An unbalanced focus on the simple problem of impeded mobility – traffic congestion – deflects attention from more important challenges facing transport in the twenty-first century: for instance climate change, oil depletion and the increasing cost to human life and health of private motor vehicles on public roads. A better term for this enlarged scope is 'integrated transport': transport that serves the wider human purposes of the city. 'Sustainability', strictly defined, is one of these wider human purposes.

In this chapter we first review the challenges of transport: what ought to be considered part of an integrated transport system, and what lessons can be drawn from existing transport practices and technologies that show the way to integrated transport. It is not difficult to find examples from around the world. The problem, however, is how to forge from these examples a new mainstream transport planning practice. Progress to this end is at best slow and may easily be diverted into channels that return to simplistic thinking. What then is slowing or stopping integrated and sustainable transport? The problem is not lack of policy but failure to implement the spirit of policy: integrated transport.

As Banister (2005: 93) points out, 'the role of institutions is crucial to the effective implementation of challenging options on sustainable development'; but time and again it appears that institutions block the way. We explain what is meant by 'institutions'. Then we consider a paradox. On the one hand the public looks to the more responsive organs of governance – liberal democracy and the market – for solutions to new and emerging problems such as climate change. But these organs of governance are designed to embody a short term perspective. The future is heavily discounted and tends to become someone else's problem. On the other hand the more enduring institutions of governance, the bureaucratic apparatus of the state, and the nexus of private interests surrounding it – less visible to research and less accountable to the public – is capable of carrying forward a long term perspective, but is also less responsive to new needs for action. It is the enduring institutions of governance that have to change – both the organizations and the people within them. This change needs to happen with unprecedented speed. But for that to occur, institutional behaviour has to be better understood.

Challenges of Sustainable Transport

There are two ways in which the word 'sustainable' is used: to refer accurately to an external reality, and as a political trope. The latter is by far the most common. A trope is a word or phrase designed to trigger an emotion, to make us feel that something is either good or bad, that something is to love or to hate, to work for or against. A classic recent example is 'the war on terror'. 'Terror' is bad, and therefore justifies war: righteous war. 'Sustainable' is likewise a word used to justify all sorts of policies and projects. Tropes like this conceal a wealth of complexities and conflicts. Beneath them there is always a reality, but not necessarily the reality the user of the trope wants to suggest.

If we wish to use 'sustainability' to reference reality then we must narrow the focus. The word was originally used in the phrase 'sustainable development'. Back in the early 1970s environmental scholars were concerned that unlimited resource exploitation by global capitalism could not be sustained without irreversibly damaging the planet's critical ecosystems (Ehrlich, 1971; Meadows et al., 1972). There were, they claimed, 'limits to growth'. But the global spread of economic growth was also gradually delivering a better and longer life to the people of the poor nations of the world, or at least promised to do so (however, see Meadows et al., 2005). So there was a conflict between ecological conservation and global economic growth. A form of growth, or 'development' as it was more politely named, had to be discovered that did not lead to the destruction of the planet's ecologies. Thus was born the trope, 'sustainable development'.

The environmental effect of unlimited economic growth is now apparent in the reality, scientifically explored, analysed and now well understood, of climate change (Solomon et al., 2007). Actually 'climate change' itself is a completely inadequate term to describe the total change in the biosphere that climate science is starting to reveal. Whereas climate change merely suggests that some places on Earth are going to get warmer and others may get cooler, the reality is that the process of global warming triggered by excess carbon emissions is tipping the biosphere into a radical change of state (this is the picture Lynas, 2008, draws from a review of hundreds of reputable scientific journal articles). Not only the Earth's atmosphere, but also its waters and inhabited land surfaces will change in a way that will affect all life. On the basis of what we know of similar changes of state in the geological past, the composition of species on Earth in a new state of the biosphere will take a very different form from that of today (Ward, 2007). It is an open question whether human life can be part of this new composition.

The world is heading for an atmospheric concentration of CO_2 of 800 parts per million by the end of the century. The last time that level occurred was in the Eocene period. So Earth's destination on current settings is a 'new old world' of 60 million years ago. As Ward (2007: 167) points out, 'the fastest natural changes [in CO_2 levels] that we are sure about are those occurring at the ends of the recent ice ages, when carbon dioxide levels increased about 80 parts per million in the space of 6000 years. That rate is about one hundredth that of the changes currently occurring. Our world is hurtling toward carbon dioxide levels not seen since the Eocene epoch 60 million years ago, which, importantly enough, occurred right after a greenhouse extinction'. 'Sustainable' is anything that will avoid that horrifying destiny and keep us in the present (Holocene) state of the biosphere.

If the world is to reach a sustainable state which is safe for life on Earth in the long term, getting to that state will have to be conducted under a global agreement among the nations, and a regime to implement and enforce that agreement. The UN Framework Convention on Climate Change is that agreement, and the Kyoto Protocol is its most recent implementation device (UNFCCC, 2011). But, as is well understood, the Protocol is inadequate because it does not mandate emission levels for the developing nations and new industrial economies. To do so would be unjust. The developing nations and new industrial economies did not overload the atmosphere with carbon in the first place, and their per capita emissions mostly remain far below those of the developed world. Article 1 of the Universal Declaration of Human Rights enacted by the United Nations on December 10th 1948 states: 'All human beings are born free and equal in dignity and rights'. Specific rights to the environment in this sense were not mentioned in the Declaration because the Declaration preceded general awareness of what was at stake. But there can be little doubt that the proposition that some people have more right to pollute than others in pursuit of economic growth would contravene Article 1 (see also the discussion by Baer et al., 2000). Therefore we can reasonably say that a minimal and anthropocentric principle of environmental justice is that each human has an equal right to use the common pool resource of the atmosphere in pursuit of economic growth. Let us call this the first principle of environmental justice. A principle that addresses the rights of a single species (humans) by no means embodies environmental justice in a full and inclusive sense, but it is at least a necessary precondition for a truly global agreement on climate change.

For all practical purposes, therefore, 'a sustainable transport system' can be defined as a system that provides the highest degree of mobility for a particular place consistent with the maintenance of a stable global climate under the first principle of environmental justice.

Overridingly important though it is, 'sustainability' is not the only challenge confronting urban transport. A second challenge comes from the loss of life, injury, and illness from road use. Globally, road traffic accidents claim a staggering 1.2 million lives annually (estimates range from 750,000 to 1.8 million), with perhaps as many as 20–50 million injuries (WHO, 2004). This figure elevates transport mortality and morbidity to the status of a major global health risk. Some researchers have characterized this condition as a 'pandemic' and a 'neglected epidemic' (Mohan, 2003; Nantulya and Reich, 2002). This mass death and injury may well be sustainable – indeed it is being sustained – but it is no more acceptable on humanitarian grounds than AIDS or malaria, particularly as those who suffer the trauma are often not those who cause it.

Socio-economic status is implicated in international and national accounts. Low and middle income countries account for 90% of traffic injury-related deaths, occurring at the rate of 20.2 per 100,000 of population. Whereas in high income nations these deaths occurred at 12.6 per 100,000 of population (WHO, 2004: Table 2.1). If the West exports car industries and the pursuit of automobility to poor countries, it is also exporting road trauma. Distribution of trauma is highly skewed. For high-income nations, most deaths involve car occupants, but road crashes in poorer nations are more likely to involve the transport modes of the poor: crashes involving two wheeled vehicles, and vehicle-pedestrian crashes (WHO, 2004).

While accident rates have fallen in the OECD, some predict that growing global wealth will increase road accidents. Kopits and Cropper (2003) demonstrate a historical link between rising income and road fatalities until a higher income level is achieved (around US$8600 in 1985 value), when the association is reversed. Hence, rising world income leads to speculation that the global road death toll will increase by 66% between 2000 and 2020 (Kopits and Cropper, 2003).

Individuals and communities subject to the influence of dangerous emissions to air from transport sources suffer increased mortality and morbidity, a problem likely to increase with the trend of urbanization. For instance, Teufel et al. (1999) concluded that in Germany cars were responsible for some 47,000 deaths annually, with many millions being affected adversely by major or minor illnesses resulting from inhaling particulates and gases. Air pollution health effects have been subject to much research, and health impacts are typically in proportion to exposure levels, where thresholds for effects are absent or very low. Road traffic is, and is likely to remain, a major source of gaseous and suspended particulate urban pollutants, especially in lower income nations. For example, road transport is the major source of nitrous dioxide and benzene in urban areas and a significant source of fine and coarse particulate matter. Of particular concern are those traffic corridors where these pollutants become concentrated, remembering that in cities outdoor air quality is a determinant of indoor air quality.

Ecological sustainability and public health are indisputable public goods which should be available to everyone. If they are threatened by the transport system, then the threat should be removed. More contentious are such matters as the distribution of transport opportunities and the quality of urban space. Urban systems are markets in which the commodity being traded is something extremely elusive: quality of place, or as real estate agents say, 'location, location, location'. The availability of public transport is undoubtedly one aspect of quality of place, but there are many other aspects including space in and around the home, proximity of suitable job and recreational opportunities, accessibility to high quality schools, proximity to like-minded people who may become friends, and local environmental quality. This last aspect can only be defined subjectively, even though there can be a degree of inter-subjective agreement. While most people can agree on what constitutes a truly horrible local environment, there seems to be room for infinite disagreement on what is a beautiful environment, and even on what it is about a particular environment that makes it horrible, tolerable or beautiful. Moreover, what people like is liable to change with their life stage, with fashion, what they are used to, and what they can imagine. For example at one time many people in the developed world seemed to place high value on home ownership, a garden, a garage and a car, all set in streets with trees and surrounded by greenery. We could call this the leafy suburb ideal. At very low residential densities together with sparsely located non-residential development, such places are difficult to serve adequately with public transport. In recent years this ideal seems to have been overlaid by the value of access to opportunities such as coffee bars and restaurants, shops and street life – the sort of environment people may typically encounter in historic city centres in Europe: the vibrant urban lifestyle ideal. Such urban places supporting higher residential densities and a higher intensity of non-residential uses are much easier to serve with public transport. The virtue of an urban market is that it allows people to choose the particular mix of qualities they want, and to trade this mix off against what they can afford. It has always been the case that cheaper places have less to offer in quality of urban space. That is why they are cheap. But should cheap housing necessarily be equated with poor or expensive transport?

We now come to a crucial question: should a public transport system provide high levels of city-wide accessibility to all places in an urban system – of high, medium and low residential density? The answer depends on the political culture of the society that hosts the urban system in question. In a society with a deep seated belief that the market rewards individual virtue and should be the dominant distributional force the answer will be that people only deserve what they can pay for. Therefore public transport that provides a high level of accessibility should only be provided where it is most cost effective, i.e. in areas where there is a greater intensity of activity, usually by virtue of employment-based development. If people live in areas of 'public transport poverty' that is their choice. But in a society with a deep seated belief in the collective, be it expressed in the liveability of the metropolitan city, the spirit of egalitarianism, or planning in 'the public interest', the answer will be different. In this perspective, accessibility is also a public good that should be available to all. While recognizing that an equal service quality cannot be provided throughout the urban area, the goal under this belief is to provide the best practicable level of service from all residential areas of the city to all centres of activity (whether local or central) in the urban system.

Fortunately perhaps, very few societies today correspond with either the pure market or the pure collective ideal. In the twenty-first century individual choice is reconciled in particular ways with the collective interest. But there can be a bias one way or the other. For example, Melbourne in Australia expresses a bias towards market distribution. Outer suburban residents are expected to provide their own personal (car) transport, paradoxically with vast public expenditure on road infrastructure. Zürich in Switzerland displays a more collective bias. Outer suburban residents in the many villages of the Canton are provided with 'best practice' public transport. The authors' position in this book is that a movement towards a more collectively beneficial outcome is justified and desired by city populations in the developed world, even in market oriented societies such as those of North America and Australia. The public goods of sustainability and health require it, and the more contested public good of accessibility cannot be achieved without it. But this movement is blocked by society's institutions.

The Role of Institutions

The political sciences try to explain why human collective action turns out the way it does. They bring into focus ideas such as society and social systems (for example 'capitalism'), power, politics, markets, the state and bureaucracy. They provide, 'valuable arrangements of true information' in the words of Stretton (1969: 426). He continues, 'The mixtures of fact, value, selection and reasoning will vary with the problems. But a non-valuing, laws-only program is simply mad' (ibid.). Our values drive us to seek an explanation for the perpetuation of unsustainable and dangerous transport systems, perpetuation not in the simple sense that constructed infrastructure remains in place and is not being torn down, but that the process of construction of these systems and their infrastructure continues in the face of an unfolding environmental catastrophe to which they contribute.

Transport systems are not simply technical machines that appear on the urban scene in automatic response to problems of mobility. They are also political artefacts – they have to be built, and decisions have to be constantly made to build them. These decisions reside at the political level, mobilizing the resources of the state. One kind of explanation that corresponds with a popular conception of democracy is that politicians make decisions in response to public demand, just as firms make decisions in the market place in response to demand. If they don't respond to demand politicians are voted out, and firms go out of business. Fortunately, there is some truth in this view of the democratic political world. But it is not the whole truth. The counterpart of this explanation from those who don't like the policies on offer is that politicians respond to and conspire with powerful vested interests. Either way the explanation is at the surface of day to day politics: the 'play of power' (Lindblom, 1968). But if the play of power is a kind of game, it is not one that is ungoverned. What governs the play are the institutions of the society in question – whether that society is governed by democratic institutions or institutions of any other kind.

In the literature of the political sciences there are different conceptions of what an institution is, and no definitive agreement among analysts (Peters, 1999: 53). However central to all definitions is the concept of rules: the rules of the political game, whether those rules are embedded in formal political constitutions (Buchanan and Tullock, 1962), constitutive of organizations (March and Simon, 1958) or embodied in belief systems and norms of individual behaviour (Denzau and North, 1994).

In the domain of transport policy the institutional dimension of policy formation takes a variety of forms. For instance, the government attitude to transport is influenced by how much it accepts neo-liberal or socialist (or social democratic) beliefs about how governments should behave and how much intervention these beliefs decree that there should be in markets. In some analyses this normative system becomes formalized as a 'mode of regulation' (Aglietta, 1979). Underlying such a mode of regulation is the real economic power of corporations manifested in control over large scale employment and investment. In these analyses a deep political structure is postulated which is devoted to managing the fundamental conflicts in capitalism between the need to maintain a peaceful and cost-effective labour force and the need to facilitate capital accumulation in private hands.

At an intermediate level between day to day politics and deep structure, the capacity of an organization to plan transport infrastructure is determined both by the financial resources to support such planning activity, and the routines accepted as justifying need. The organizational contribution to policy making is governed by the organization's culture: 'how we do things here'. Routines such as the 'four step model' provide a politically acceptable method of justifying expenditure on new roads (see Meyer and Miller, 2000; Flyvbjerg, Holm and Buhl, 2006). Skills are built around such routines and embedded in organizations. Beliefs about what the effect of infrastructure building will be are important: for instance that building infrastructure will provide employment and deliver economic growth.

All these institutional dimensions of the polity frame the way transport is delivered. Marx and Weber, among the founders of the political sciences, conceived of these structural or institutional forces that framed surface politics. Weber observed:

> Under normal conditions the power position of a fully developed bureaucracy is always overtowering. The 'political master', finds himself in the position of the dilettante who stands opposite the 'expert', facing the trained official who stands within the management of administration. (1946: 232)

The 'experts' set the choices before politicians, and all too often these choices are severely limited, or biased to steer the decision to the one favoured by the administration. Politicians don't necessarily want a lot of choice, most of the time they want to know what is guaranteed to work to solve the problems conceived by the experts. Nevertheless Weber also believed that politicians are ultimately answerable to an ethic of responsibility – 'The honour of the political leader, of the leading statesman, however, lies precisely in an exclusive personal responsibility for what he does, a responsibility he cannot and must not reject or transfer' (Weber, [1919] 1946: 95). So from time to time ethical political leaders consult their convictions and, if the current course does not serve them, order a change of course in public policy.

In a seminal essay on policy planning, the American political scientist Charles Lindblom recognized a different aspect of institutions in determining policy choices when he wrote, from the viewpoint of Weber's 'expert':

> Most of us approach policy problems within a framework given by our view of a chain of successive policy choices made up to the present. One's thinking

> about appropriate policies with respect, say, to urban traffic control is greatly influenced by one's knowledge of the incremental steps taken up to the present. An administrator enjoys an intimate knowledge of his past sequences that 'outsiders' do not share, and his thinking and that of the 'outsider' will consequently be different in ways that may puzzle both. (Lindblom, 1959: 87)

Lindblom termed this phenomenon 'incrementalism'. There are two important points here. The first is that experts and outsiders, such as public transport advocates, misunderstand why each behaves the way they do. This can lead to the alienation of outside experts from inside experts, and a good deal of friction and frustration on both sides. The more important point is that understanding policy requires an understanding of its historical development: in the words of Paul Pierson (2004) 'politics in time'.

In summary then, institutions create the framework of rules for public policy making. Institutions both limit intentional action, and make action possible by providing definitions of problems, solutions to those problems, the knowledge to implement those solutions and a corps of personnel bearing that knowledge. Institutions are 'the formal or informal procedures, routines, norms and conventions embedded in the organizational structure of the polity or political economy' (Hall and Taylor, 1996: 938). Institutions both limit intentional action, and make action possible by providing definitions of problems, solutions to those problems, the knowledge to implement those solutions and a corps of personnel bearing that knowledge. None of this complex of people and ideas is brought instantly into being, and once brought into being it is not lightly or rapidly discarded. Institutions comprise sets of formal rules that create and empower organizations – an organizational domain, and sets of beliefs and values that shape policy and the process of public choice – a discursive domain. Understanding these institutional domains cannot be obtained through a snapshot at any one time, but requires analysis of the evolution of institutions over a significant period, noting occasions when politicians have mobilized a major institutional or policy shift.

Historical Institutionalism in Transport Policy Analysis

Historical institutionalism 'situates the analysis of institutions within the structural development of the state as it, or more accurately its parts, respond to the conflict among powerful social actors, so as to privilege some interests while demobilizing others' (Hall and Taylor, 1996: 937). The new institutionalism also pays attention to the informal aspects of institutional life, particularly networks of actors and their cultural norms subsisting within and between formal hierarchies (Torfing, 2001: 281).

Exemplifying the new institutionalism in the transport policy field, Dudley and Richardson (2000) and Vigar (2002) have developed powerful analytical frameworks to explain, in institutional terms, policy change and continuity in the British post-war transport domain. Dudley and Richardson note the 'inherent paradox' in explanations of change in systems 'which exhibit innate qualities of stability' (ibid.: 229). Transport policy typically exhibits those innate qualities. For long periods a particular paradigm prevails in which a solution (for instance road building) is used to address a problem (traffic congestion) with a variety of policy technologies (travel demand projection and cost-benefit analysis based on travel time saved) supported by well-funded and powerful organizations (roads authorities or Departments of Transport) and supporting networks of people (consultant engineers, bureaucrats, lobbyists, industrialists) acting in concert.

In Britain, 'Prior to the mid-1950s', Dudley and Richardson (ibid.: 70) observe, 'the road lobby was fragmented, and consisted of epistemic communities of civil engineers, together with a variety of motoring and lobbying organizations ... but for many years it could not be said that they constituted an effective advocacy coalition, i.e. an aggregation of actors composed of people from various governmental and private organizations who share a set of normative and causal beliefs and who often act in concert' (Sabatier, 1993: 18). Similarly in Australia up to the 1950s Davison presents a story of passionate advocacy of the car on the part of motoring organizations, journalists and a few politicians at a time when 'the motor car is, as yet, a rich man's hobby' (Davison ibid.: 8, citing a government publication: the Queensland Yearbook, 1945). But it took a confluence of values between 'the freedom of the road' and 'individual liberty' to propel the car-road solution to a hegemonic position. Davison describes the growing alignment of the emergent road lobby with the values espoused by the post-war Liberal federal government led by Robert Menzies. In the early 1950s federal political values conflicted with the support for public transport from socialist ideology and state governments, 'which, as the proprietors of publicly owned transport systems, were reluctant to allow cars and motor buses and motor trucks to drive their enterprises out of business' (ibid.: 125).

Dudley and Richardson show how a 'third order change' was accomplished in Britain to move the road building solution to a position of hegemony. The concept of 'third order change' comes from Hall (1993). If 'first' and 'second' order changes describe normal policy making, the play of power amongst policy actors and incremental change, third order change refers to periodic discontinuities in policy (ibid.: 279), events in which not only policies but the institutional base of policy making starts to shift. Thus Dudley and Richardson refer to Kingdon's conception of three 'policy streams' flowing through the system: problems, policies and politics (Kingdon, 1995): 'although these streams are separate and have their own dynamics, at some critical juncture they are joined, and the greatest policy changes grow out of that coupling' (Dudley and Richardson, 2000: 28).

What then caused the change and subsequent establishment of hegemony? Plainly, as car industries producing affordable cars coincided with growing personal incomes in the 1950s and '60s in both Britain and Australia, growth in car ownership produced the very visible 'problem stream' of traffic congestion. The solution, large scale expenditure on roads in both Britain and Australia, was initially blocked by Treasury departments. In the middle to late 1950s in Britain, however, a succession of ministers 'took a personal interest in not only developing the knowledge and expertise of the epistemic community of highway engineers, but also in ensuring that road building became the dominant policy "frame" within the Ministry of Transport generally' (ibid.: 71). The Minister, Dudley and Richardson argue, 'has the authority and means to transform a collection of advocacy groups into an effective advocacy coalition' (ibid.). However, if ministerial action was necessary to propel the idea that 'there are votes in roads', and to maintain that momentum, there was also a feedback effect:

> As road construction grew, so there was a natural process whereby the epistemic communities of road builders became a dominant force within the MoT [Ministry of Transport]... Once the political importance of road building had been established, then a process of feedback could take place whereby a hegemonic policy community could amplify the roads message over time and across many policy arenas. (Dudley and Richardson, 2000: 109)

In Australia, Davison, as we have noted, commented on the support of the Prime Minister (Menzies) for private car ownership at a critical period in the post-war years. At state level, inspired by American models of highway planning, the road lobby coalesced around transportation plans for the major cities. Just as British motorway plans were influenced by the German autobahns, so Australian highway plans were influenced by American thinking. Scholarships were established to send Australian engineering students to Yale University to take the Traffic Engineering Course. In the state of Victoria the government, led by a State Premier who strongly supported major expenditure on motorway building (Henry Bolte), legislated to create the Metropolitan Transportation Committee, 'an unwieldy amalgam of the many competing state and local governments and special-purpose statutory authorities' (Davison, 2004: 178). Unwieldy it may have been, but this committee provided the vehicle for the increased influence and eventual hegemony of the road engineering epistemic community.

A second 'third order' change began in Britain in the 1970s when the rationale for the roads program, driven by popular consumerism, was brought into question by the rise of environmentalism, the oil shocks, and the need for budget cuts. Once again, Dudley and Richardson (2000) point out, the change was orchestrated by a Minister of Transport. 'Environmental ideas rode in on the back of exogenous factors such as public expenditure crises and the steep rise in oil prices, but from 1976 the dynamics of policy change took on a more endogenous character' (ibid.: 139). The road building hegemony, however, re-established itself under the Thatcher government's 'Roads for Prosperity' program until a further round of third order change took place towards the end of the 1990s.

The 'new realism' may have been proclaimed by Phil Goodwin et al. (1991) and adopted under the rubric of 'sustainable transport'. But elements of the roads hegemony linger on. Doubts about the sustainability and strength of the third order change remain. Dudley and Richardson remind us at the end of their study, that the [New Labour] government might be "anti-new roads" but it was fearful of being labelled as "anti-car"' (Dudley and Richardson, 2000: 196). Publishing two years after Dudley and Richardson, Vigar (2002: 185), who also investigated the change to the 'new realism', noted that although the 'predict and provide' paradigm appeared to have run its course, 'what is less clear is what emerges in its place as the "new realist" approach has yet to cohere into anything that can resemble a paradigm and therefore have the authority to resist a potential plethora of incremental policy incursions such as roads programme expansion.' Dudley and Richardson (2000: 191) argue that the change in the later 1990s was real, and not just a shift in rhetoric: they say that in the 1996 budget 110 road schemes were axed worth about 6 billion pounds. Figures for road construction to some degree support the contention. In the years 1997–98 and 1998–99 133 and 96 kilometres respectively of new motorways and all-purpose trunk roads were completed. Between 1999–2000 and 2007–08 motorway and trunk road completions averaged 47 kilometres per year (Department of Transport UK, 2007, Table 7.16). There was a significant change. Yet in terms of the change really needed to address the transport challenges discussed above, 'paradigm shift' may be too strong a word. Between 1990 and 1997 in the UK the mode share for travel by cars, vans and taxis rose from 85% to 87% and fell thereafter only slightly to 84% in 2007. The mode share for rail travel was stable at about 6% rising to 7% between 2005 and 2007 (Department of Transport, 2008, Table 1.1 Passenger Transport by Mode 1952–2007). There was little reflection in travel behaviour of any supposed paradigm shift in policy settings.

A somewhat parallel process of deeper change has taken place in Australia. The 1970s saw the first shift in the road building hegemony. The exogenous factors of the oil shocks, the rise of environmentalism and citizen protest (backed briefly by a powerful construction union – the Builders' Labourers' Federation) against the building of motorways in inner urban locations (Sydney, Melbourne and Perth) found support at ministerial level, both in the federal government and the States. However, with the acceptance of tunnelling technology for driving roads under environmentally sensitive locations, the adoption of private-public partnerships to provide additional finance, the availability of space in the outer suburbs to continue the road building program and the consolidation of expertise and authority in transport departments has ensured the continued survival of the roads hegemony. The 'empty chair' of public opinion was filled by 'suburban man' who was allegedly wedded to his car. The Australian case is discussed in more detail in subsequent chapters.

These historical analyses deploy a variety of concepts to capture the nature of institutional behaviour, for instance: advocacy coalitions (Sabatier, 1988; Sabatier and Jenkins-Smith, 1993), third order change (Hall, 1993: 279), policy viruses (Dudley and Richardson, 2000: 15), the role of key individuals (Bevir and Rhodes, 1999: 225; Dudley and Richardson, 1999), the 'empty chair' principle (Dudley and Richardson, 2000: 84), the 'hollow core' (Heinz, et al.,

1993), and time scale and timing (Pierson, 2004, Chapter 2: 54–78). Four more general points may be made about such institutional analysis. First, whilst the importance of institutions has long been recognized, institutions have mostly been treated as contexts for the investigation of public policy making rather than the object of investigation itself: in gestalt terms, the ground rather than the figure. As Pierson remarks, 'Analysts working from a variety of perspectives have produced compelling work on the significance of institutional arrangements for political and social outcomes ... By contrast, we have made less progress in treating institutions as themselves important objects of explanation' (Pierson, 2000: 475). Second, once institutional change is brought to the foreground, time becomes an important variable: not only how but when events occur, and in what order – what comes before what. A snapshot view is thus inadequate to describe institutional behaviour. Even though a conjunction of political forces may trigger a rapid policy shift, the policy paths towards that event may be many years in gestation. Thirdly, individuals in positions of political power, typically government ministers, play a central role in institutional change. In all the cases which have become examples of 'best practice' transport and land use planning, a political leader has mobilized the change from a more typical or normal policy pattern: Jaime Lerner in Curitiba, Enrique Penalosa in Bogota, Michael Harcourt in Vancouver, Hans Künzi in Zürich, Bill Rodgers and Brian Mawhinney in the UK. Finally, while institutional change is of great interest, institutional continuity appears to be more the normal state and in need of explanation.

Discussing the human mechanism for perceiving small units of time, Klein (2006: 44) asks: 'But why did nature rely on pacemakers that control the movements of our muscles, rather than giving us our own sense of time?' He answers his question: 'life forms did not develop the way ingenious engineers might have created them. Evolution is conservative. Once it has found a solution to a problem, it generally sticks with it. The blueprint is passed down from one generation to the next and from one species to the next, and modified only minimally when new tasks enter the picture. As long as an existing principle continues to suit the purpose reasonably well, it is retained, even if it is barely functional and a new construction would be far more suitable.'

Conclusion

The world can make slow progress towards improved outcomes from transport systems: the value of individual personal mobility can be preserved with fewer people killed or maimed by road vehicles, fewer people with health damaged by air pollution, reduced local impacts on the habitats of endangered species, better management of public transport systems, more humane and attractive urban environments through which transport channels pass. Steps in this direction can be incremental, and the risk of inaction is no greater than those which societies have tolerated up to now. But effectively containing climate change requires institutional change of a quite different order. The long term risks of climate change are inordinately greater than any risk human society has ever before confronted. The tipping points, beyond which catastrophic climate change is locked in, are fast approaching. By contrast with the change required, between 70 and 90% reduction of GHG emissions, the third order change in transport recorded by institutional analysts to date appears as merely a somewhat larger incremental step.

The scale of institutional change required is much more like that which occurred during the Second World War: a rapid and radical mobilization to fight a clear and present danger (Gilding, 2011: 131). Change must also take place within each nation under the rubric of a global perspective of the problem and of just solutions. First, though, we need a coherent explanation of institutional momentum and continuity, a perspective that brings together the concepts of institutional behaviour in a logical framework. The contention of this book is that 'path dependence' provides such a framework, and we explore this framework in the next chapter and how it might apply in the case of urban transport.

Chapter 3
The Irrationality of Path Dependence

Introduction

Political institutions confront us with a dilemma. On the one hand they have the virtue of endurance. On the other hand they must change in response to changes in their social environment. Democratic institutions resolve this dilemma by building in responsiveness to the public they serve. Even so, public policy normally changes only at the margin, leaving the bulk of its settings untouched. Today this dilemma is compounded by the need for institutions to respond to changes in the global geophysical environment whose effects on society can only be anticipated in terms of risk (Stern, 2007; Garnaut, 2008). Action, once the effects of global warming and fuel shortage are strongly felt by societies, will almost certainly come too late. Transport policy demonstrates this compound dilemma.

It is important to develop a coherent understanding of why policy endures beyond its rational basis, and how the policy system could become more responsive to change in its geophysical context. The concept of path dependence provides such an understanding. Path dependence is an irrationality of governance, a governance *failure*, when public policy ceases to coincide with the public interest and is instead determined by the force of its own inertia. This failure becomes catastrophic when the threat is as severe as global climate change.

The Theory of Path Dependence

Simply stated, the theory of path dependence posits that the production of commodities does not entirely follow the rational logic of the market in which supply is led by demand. Rather, chance events play a pivotal role in deciding which kinds of product are taken up and become dominant in the market. And, once established, there are incentives within real production systems to keep producing the same kind of product. Similarly within politics, the production of policies and institutions does not entirely follow the democratic logic of response to the public demand. Rather, chance events determine which policies and institutions become dominant and there are incentives within real political systems to maintain established policy settings.

The theory therefore makes two main claims:

1. The presence of a product or policy at any one time can only be properly explained by examining its historical development. In certain circumstances

the confluence of events at the time of initiation of production (of product or policy) determines the path taken that leads to the present situation.
2. For some products and policies (but not all). There is increasing value to the producers to keep producing more of the same rather than change to a superior alternative.

Our starting point, then, is the matter of time and timing. We then need to consider the concept of 'increasing returns'. We start where the theory starts with the production of commodities, then we move to the application of path dependence in the politico-institutional system, and finally to the transport system.

History, Time, and Timing

Economist Steve Keen (2001: 79) points out that, 'the absence of time in its analysis – is one of the fundamental weaknesses of conventional economics'. Conventional economists will deny this and say that indeed 'time' is one of the most important variables in their explanatory scheme. It is, but the economists' 'time' is time that stands still. It is packaged in small quanta of 'change', which is very different from real historical time during which firms develop and many things happen that are extraneous to the pure logic of supply and demand. Brian Arthur in the 1970s claimed that in the historical process of the evolution of business, other things happen that are not consistent with the static rationality of conventional economics.

Increasing Returns

Some things that could happen in the course of a firm's development upset the normal assumptions of rational equilibrium in the economy. Conventional economics posits that production of each commodity will reach an equilibrium point truly reflecting the effective demand for the product. The profit curve will gradually decline as more and more of the commodity is produced and more and more competitors enter the market. This continues until any extra investment in the production of the commodity reaches zero and the commodity will be produced at the lowest price to consumers. Thus market discipline was applied to the behaviour of firms. Arthur proposed that, on the contrary, in certain cases self-reinforcing factors exist in the logic of production to induce *increasing* returns to further investment in the same product. Increasing returns become possible when a firm is able to gain market dominance, which in turn allows the delivery of monopoly profits.

Features of path dependent processes are that: a) early potentially random events have a large impact on later outcomes, b) the path becomes inflexible as it unfolds over time and it becomes harder to shift from one path to another, c) accidental events early in the development process cannot be ignored, because they feed back into future choices, and d) in the long run the outcome that becomes 'locked in' may generate less than optimal, or negative, outcomes (Arthur, 1994: 112–113; Pierson, 2000: 253).

Not all economic processes are subject to increasing returns and path dependence. But Arthur (1994) argues that four factors predispose a production technology to increasing returns:

- *Large set-up or fixed costs.* If the initial investment required to create a production technology is large, these 'fixed costs' per unit produced can be reduced by spreading them over a larger and larger output from the same technology. Total fixed cost per unit falls as output increases, and there is therefore an incentive to stay with the technology used at the start.
- *Learning effects.* Over the course of time in the use of a technology, firms learn better ways of producing the same goods and at lower cost. Consumers learn to use products more effectively, which may stimulate innovations in the same product line. So again, there is an incentive to stay broadly within the technology.
- *Coordination effects.* These occur 'when the benefits an individual receives from a particular activity increase as others adopt the same option' (Pierson 2000: 254). They are particularly significant when a technology has to be matched to a particular infrastructure. For instance the use of private vehicles requires an infrastructure of roads, fuel stations, repair workshops, on-road breakdown services, traffic control measures and so on. The infrastructure, itself usually requiring large investment, and the numbers of users creates a momentum to continue with the same technology.
- *Adaptive expectations.* As a type of product increasingly prevails on the market, the belief grows that it will continue to prevail in future (Arthur, 1988a: 10). 'Projections about future aggregate use patterns lead individuals to adapt their actions in ways that help make those expectations come true' (Pierson, 2000: 254).

In a well-known critique of increasing returns, Liebowitz and Margolis (1990), show that some of the key empirical examples are open to challenge (The Qwerty keyboard, the Betamax format video recorder). These authors (1995) later mounted a more theoretical defence of the rationality of markets arguing that only 'remediable' path dependence is of theoretical significance: 'That is, there exists, or existed, some feasible arrangement for recognizing and achieving a preferred outcome, but that outcome is not obtained' (Margolis and Lieberman, 1995: 207). Pierson (2000: 257) however points out that the argument mounted by Liebowitz and Margolis has little relevance for institutions. They simply assume that where step by step improvement (incrementalism) fails to deliver improvements, institutional innovation will do so: (Liebowitz and Margolis, 1995: 224). But regulatory institutions did not arise that could have prevented a series of economic crises occurring in the USA culminating in the global financial crisis. The point here is that regulatory institutions are politically created (and destroyed) and are subject to path dependence so that it is difficult to make the necessary basic changes.

From the point of view of transport systems, in this book we are more interested in transport institutions than the behaviour of firms. It is not that the behaviour of individuals and firms as users of transport systems and producers of vehicles is irrelevant, but the leading role in shaping urban transport systems is overwhelmingly played by governments. Urban transport systems are primarily created by planning, investment, and management of physical infrastructure: roads and railways, ports and airports. A varying structure of public sector organizations is always involved in managing public transport and regulating road use. And these infrastructural investments and organizational systems are primarily those of governments determined by political institutions. So in considering path dependence we now need to turn from the economy to the polity.

From the Economy to the Polity

Pierson (2000: 257) shows how the theory of path dependence can be applied to understanding political behaviour. He discusses four interconnected characteristics of politics 'that make this realm of social life conducive to increasing returns processes': the central role of collective action, the high density of institutions, the potential for using political authority to enhance asymmetries of power, and the intrinsic complexity and opacity of political systems. These characteristics typically come into play in the development of transport systems. The travel behaviour of consumers of transport may be described as the aggregate of their many individual choices. But these choices are conditioned at any one time by the opportunities provided by the transport infrastructure and its management provided collectively. If the supply of public transport is poor or non-existent, consumers will not (or cannot) choose that option; similarly with cycling or walking if these options are seen to be unsafe or unattractive. There is often an assumption, especially on the part of Treasury departments, that infrastructure should only be provided where demand is demonstrated, on the assumption that demand is independent of supply (of transport opportunities). This accounts for the popularity of the technology of 'predict and provide'. 'Predict and provide' coupled with the assumption that demand for transport modes is independent of supply produces transport system path dependence because of the circularity:

- Transport infrastructure influences mode choice,
- Past mode choice influences future supply of infrastructure.

While 'predict and provide' may have been discredited in the scholarly discourse of transport planning (Owens, 1995; Goodwin, 1997), it continues to be influential in transport planning practice.

Provision of transport infrastructure requires organizations, norms and rules – institutions. As Pierson remarks, 'efforts to coordinate actors in the pursuit of public goods often require the construction of formal institutions'. 'For my purposes', he writes, 'the fundamental feature of politics is its preoccupation with the provision of public goods' (Pierson, 2000: 257). Transport systems have so many of the features of public goods that, even if certain elements may be provided under contract by private constructors and operators, and even if users may be made to pay for use of some of those elements (road pricing via tolls, and public transport fares) a transport system as a whole must be considered a public good. Politics produces policy and institutions which, if they have impact at all on the external world, do so through legally binding rules: 'Legally binding rules are not just a foundation for political activity (like property rights in the economy). They are instead the very essence of politics' (ibid.): Pierson's third and fourth characteristics of politics, to which path dependence is relevant, are the potential for power asymmetries and the opacity of politics. In the first instance Pierson refers to occasions when power over the political agenda is hidden and uncontested through anticipated reactions and ideological manipulation. In the second, he refers to the weak potential of political systems to receive, hear and act on feedback from the public on the success or otherwise of public policies. In the transport literature there are examples of encounters with both these characteristics (see Vigar, 2002; Dudley and Richardson, 2000).

Recent work on the politics of transport focuses on the longevity and resistance to change of coalitions of actors once they become dominant. For instance, Stone (2008) in an examination of transport policy in Melbourne (Australia), Vancouver (Canada) and Perth (Australia) concludes:

> Empirical evidence confirms the theoretical understanding that a dominant urban transport policy network is very resistant to change. Theoretically some analysts seem to believe that policy network change is possible without a substantial change in membership of the dominant policy network. Empirically, however, change away from automobility seems sufficiently difficult that it does not happen without changes in policy network membership. This seems, in turn, to require change at the level of the governing coalition through which supporters of alternatives gain influential positions in a new coalition that brings together sufficient resources to enact new policy and to weaken the influence of previous approaches. (Stone, 2008: 75)

This is a form of path dependence in which coalitions of actors are able to derive value from consolidating an established policy setting and preventing alternatives reaching the political agenda where reassessment of policy might lead to change. None of this should be read as suggesting that change in policies or institutions does not occur. The fact that it does in some instances in some places (see for instance Bratzel, 1999; Stone, 2008, 2009) helps focus the question of why most of the time it does not.

Windows of Opportunity

The theme of timing and conjunction of events ('critical junctures' in the path dependence terminology) runs through the seminal work of Dudley and Richardson (2000) on the development of British transport policy. To illustrate 'critical juncture' we draw heavily on the work of these authors as they trace the history of a particularly striking example: that of the Standing Advisory Committee on Trunk Road Assessment (SACTRA) (Dudley and Richardson, 2000: 143).

During the 1970s, following rising public concern over the environmental effects of roads, building roads gradually came to be seen as 'a problem rather than a solution' (ibid.: 133). Path dependence theory signals the importance of chance events, and Dudley and Richardson note: 'In the 1970s, notwithstanding the efforts of policy entrepreneurs such as John Tyme and a range of environmental groups, there was a significant element of unpredictable chance in the manner in which environmental ideas took root in the roads policy system'. Nevertheless the growth of environmentalism focused on transport issues was one vector of influence.: Another vector was a series of public expenditure crises between 1973 and 1975. Together these led to the perception in political circles that road building had a cost both in scarce political and financial capital: a 'double window of opportunity' opened up as the authors put it, 'where exogenous and endogenous factors reacted on each other' (p. 121). The minister of the day, Bill Rodgers, was 'an important carrier of new ideas and values, transforming them into negotiated policy change' (p. 133).[1]

In 1976 Rodgers appointed a committee under the chairmanship of Sir George Leitch, a businessman, to review the department of transport's methods of assessing major road schemes and traffic forecasting. This committee, which included academics as half its membership, was distanced from the pressures of day to day politics and equipped with the legitimacy of professional norms. The committee in its report argued for shifting the evaluation of roads from simple monetary benefit-cost assessments to a comprehensive framework 'which would embrace all the factors and groups of people involved in a scheme assessment' including the more intangible environmental factors (p. 137). The Leitch report succeeded in constructing a 'settlement' between the conflicting interests of the road and environmental policy communities. SACTRA was created as a permanent successor to the Leitch Committee to pursue the Leitch Committee's reform agenda.

The contents of SACTRA's report *Trunk Roads and the Generation of Traffic* are now well known to transport planners: confirming the existence of 'induced traffic' resulting from new road construction, a finding that signified a key turning point in UK transport policy. Important here is the summary given by Dudley and Richardson (p. 181) of the window of opportunity that culminated in the SACTRA report.

1 Giving capitals to the above terms (Executive, Broker etc.) indicates that these are generic roles identified in the political science literature, for example, Headey (1974), Sabatier (1988).

> In the climate of 1989 which produced Roads for Prosperity, SACTRA's answers to the question of whether new roads generate extra traffic might have been employed as a good reason for building more roads. By late 1994, however, in a climate greatly more amenable to the principles of sustainable mobility, a positive answer to the question by SACTRA was likely to provoke a different response. (citing SACTRA 1994: para 15.03)

Subsequently the SACTRA report gave politicians ample reasons for dismantling the policy setting announced in *Roads for Prosperity* and setting forth on the path of the 'new realism' (Goodwin et al., 1991). Illustrated here is the fact that critical junctures occur not as flashes from out of the blue, but as a result of the gradual flowing together of political vectors over a significant period of time, in this case the 20 years between about 1974 and 1994.

Windows of opportunity, or 'critical junctures' stimulating shifts in policy paths occur when a political vectors flow together to create change. The history of the SACTRA report is such a striking example of a critical juncture that it is surprising that Dudley and Richardson did not make use of the path dependence framework in their analysis.

Two Aspects of Path Dependence in Transport Systems

Transport systems are complex socio-technical artifacts. They result from a mixture of individual and collective (governmental) behaviour. Transport systems are intimately connected with land use patterns and processes of urban development. From the point of view of transport systems we can usefully distinguish two aspects of path dependence: the technical and the institutional. Institutional path dependence can be further divided into two aspects: organizational and discursive.

Technical Path Dependence

As noted above, economists have pointed out the impact of increasing returns in the location of production and thus in city development (Krugman, 1991; Arthur, 1994: 49–67). Krugman (1991: 80) comments, 'If there is one area of economics in which path dependence is unmistakable, it is in economic geography – the location of production in space. The long shadow cast by history over location is apparent at all scales, from the smallest to the largest – from the cluster of costume jewellery firms in Providence [USA] to the concentration of 60 million people in the North East Corridor'.

It is not just location of production, and therefore employment, but also the physical form of cities that is path dependent. The physical form of a city can be shaped by the dependence on a particular form of transport (as first shown by Schaeffer and Sclar, 1975; and later supported by Newman and Kenworthy, 1989, 1999). In the case of Australian and most North American cities, and the suburban extensions of European cities, the widespread adoption of the private vehicle for getting around and servicing the city has made possible the choice of low density residential environments. In turn these low density environments encourage the building of high quality roads to accommodate private vehicles, thus reconciling low density with reasonable accessibility to services. Cervero (1998: 29) observes for American cities, 'The scattering of activities to all corners of the modern metropolis poses unprecedented challenges to public transportation'. In later work he argues that road investment and urban development are co-dependent:

> The causal chain works as follows: a road investment increases travel speeds and reduces travel times …; increased utility, or a lowering of "generalized cost", in turn stimulates travel, made up of multiple components, including new motorized trips (e.g. latent demand previously suppressed), redistributions (modal, route and time-of-day shifts), and over the longer term, more deeply rooted structural shifts like land use adjustments and increased vehicle ownership rates (that in turn increase trip lengths and vehicle miles travelled). While evidence on the induced-growth effects of new highways is limited, roads and prominent fixtures of America's suburban landscape – big box retail, edge cities, and campus-style executive parks – that they serve are clearly co-dependent. (Cervero, 2005: 128–129)

However, urban form (for example a dispersed city form) itself is not sufficient to explain the dominance of the 'private cars on public roads' mode of delivering urban mobility. As Mees (2000) has remarked, Toronto (Canada), which is a low density city like Melbourne, has a much more effective form of public transport. Likewise, metropolitan Zürich provides a public transport system based on the rail network that efficiently services the entire Zürich Canton comprising both a central core and widely distributed settlements whose overall density is much lower than that of metropolitan Melbourne. His point is that the lessons from the examination of 'all genuinely successful urban public transport systems, be they in Zürich, Munich, Metro Toronto or Vancouver, share a common feature, namely central, regional planning by a public agency' (ibid.: 286). Mees, then, provides an institutional, and more specifically an organizational, answer to the question of what determines the mode of provision of mobility.

Institutional Path Dependence

Hall and Taylor (1996: 938) conflate two aspects of institutions that we find useful for analytical purposes to separate: 'organizational structure' and 'routines, norms and conventions'. We might say that the former (organizational structures) are the hardware of institutions, the latter, the discourse of explanation and justification, is the software. The former are formal rules governing relationships amongst positions, setting out the skills required to fill the position and the responsibilities of those who hold them, the latter is the talk that goes on around policy, sometimes written, sometimes verbal: the storylines justifying and explaining public policy (Hajer, 1997).

The organizational aspect

The power of organizations as they develop and change over time can be an important source of path dependence. As Pierson (2004: 173) observes, 'claims about path dependence are often grounded in the scale of commitments resulting from previous institutional, organizational, or policy outcomes'. Organizations command financial resources, expertise and network support which they bring to bear on policy. Organizational strength has a self-perpetuating quality that entrenches policy doctrines about problem definition, methods of analysis and solutions. Following the work of Astle (2007) which adopts the longitudinal approach associated with historical institutionalism (Pierson, 2004), it is necessary not merely to take a snapshot of organizations at a single point (e.g. the present) but to examine their development over time. We can identify five main aspects of organizational power.

First, organizational power derives from a capacity to operate over the full span of functions relevant to their policy domain. However over time that can mean a series of organizational changes and amalgamations. How these changes are accomplished and which functions are brought within the organizational structure may either lead to internal fragmentation and conflict or to a larger consolidated scope of the organization's function (Astle, 2007: 7).

Second, organizations often function as the hub of networks of support for particular policy settings. Their capacity to form such networks, both inside and outside the state apparatus, over the range relevant to their core functions is an important aspect of their power (Stone 2009: 423; Rhodes and Marsh, 1992: 197–98). The strength of such networks has an important bearing on the ability of an organization to ensure policy continuity.

Third, an organization's power is related to its ability to influence decisions about planning, construction and management of its assets (Astle, 2007: 49). Not all of the key planning and management functions may be within the organization's formal functions. So its power derives in part from its ability to extend its influence to other domains of public policy,: for instance urban planning or treasury policies.

Fourth, an organization's power derives from both the level of expenditure and its discretionary control. For instance, if an agency channels a large amount of investment (capital) funds, it can legitimately divert some of those funds to the design of future capital projects. Agencies that are starved of investment funds will also be prevented from planning new projects consistent with their policy agenda. Also the power of an agency to decide on expenditure priorities may be both limited and ensured by the degree to which funds are hypothecated for particular purposes.

Fifth, the chain of accountability of an organization will affect its capacity to influence policy. Direct access to a Minister, for example, means that an agency can by-pass bureaucratic gate-keepers and influence the political level directly. However this relationship works both ways. A strong Minister can exert more direct control over the implementation of a policy, and much obviously depends on where policy originates: within the bureaucracy or at the political level. In the transport field, though there are notable exceptions, the former appears to be more common than the latter.

These aspects of organizational power can entrench policy positions, and, most importantly, the balance of power between organizations with different functions in the transport field can determine the shaping and maintenance of the path of transport policy.

The discursive aspect

The 'discourse network' is a concept used to explain the interconnectedness of ideas, decision-makers and their mental models of reality: what is 'important', what 'the problem' is, and how to go about solving it.

In a storyline, understandings and objective diagnoses are inseparably interwoven with values and norms. Following Hajer, a storyline is essentially an idea which can be expressed as a claim, and can then be put together with other storylines to form explanations, justifications or arguments. Storylines function to hold discourse together across the range of actors in a particular policy domain. They may also serve to position actors relative to one another. The mental models in the minds of different actors consist of varying sets of storylines, appealing to the different actors' interests and professional positions in organizational structures.

The idea of a socially constructed world view is of course not new (see for example Kelly, 1955; Young and Mills, 1980). Denzau and North acknowledge that 'many of the changes we wish to understand are social, such as changes in informal institutions or ideologies' (Denzau and North, 1994: 23fn). The idea of the discourse network suggests that the models people hold are not so much individually as inter-subjectively generated around particular policy domains within a framework provided by external (to the mind) institutions. Both discourse networks and the external institutions created to regulate the production of policy contribute to path dependence. Torfing describes the 'policy path' as 'a relatively entrenched way of unifying, organizing, and regulating a certain policy field' (Torfing, 2001: 286). North (1998) observes that 'value' is not simply to be understood in simple financial or even in quantitative terms. Value is derived from doing things in the 'right' way for the 'right' reasons. Thus he points to 'the importance of institutionally embedded mental models for shaping our perception of payoffs' (Torfing, 2001: 282).

If, as has been suggested, organizational networks are part of the external structure (that is outside the minds of actors), common networks of ideas – shared mental models – are the internal counterpart: 'To the extent that the speaker and a listener have common features in their mental models for the concepts identified in the SMM (shared mental model), they are more likely to be able to encode and decode their internal ideas into a shared language' (Denzau and North, 1994: 20). Key individuals in the policy network may talk to and persuade one another, or, in a more passive sense, their policy making activity may simply be channeled along certain lines by shared mental models, the discourse network, without much need for contact (Low, 2005).

Denzau and Grossman (1993) and Denzau and North (1994: 22–26) discuss the process of discursive change. A critically important prior distinction they make is between change within the 'model space' (learning) and change *of* the model space or 'representational redescription'. What then do they mean by 'model space'? It seems that they mean something like (but not exactly like) the notion of 'paradigm' in the history of science identified by Kuhn (1970). The dimensions of the 'model space' are shaped by one or more key assumptions that generally do not change as learning from new observational data occurs. The ideas generated in support of policy occupy the 'model space'. A useful metaphor is the shared language that develops within the model space as people talk to themselves and others about why some policy should continue to exist (or not), and write policy documents entrenching those ideas.

The Example of 'Congestion'

Though there is an obvious reality about traffic congestion which makes people angry and impatient, the term 'congestion' when it is used by transport advocates, functions as a trope. Here, 'congestion' is used to direct attention to an undesirable quality; the users mean us to think that the phenomenon to which it refers is unhealthy. The term is derived from bodily ills such as congestion of the heart or lungs. As Cresswell (2008: 131) notes, 'When, for instance, we understand that the discovery of blood circulation by William Harvey in the seventeenth century led to city planners associating circulation with health and moral order, we can better understand the urban development of cities such as Paris or Washington D.C.'. Figure 3.1 is a detail from advertisements promoting the City Link motorway enlargements and extensions in Melbourne, Australia, in the 1990s. The link between bodily health and traffic circulation is here unpleasantly explicit. We might think that this is just a bit of metaphorical fun, but the trope of congestion and its opposite pole 'healthy' circulation has a deceptive and manipulative side.

Congestion can be costed in monetary terms, and the public can be made to pay for it through a congestion tax or road pricing. Taxes of this kind have been successfully introduced in Singapore, London, and (via tolling) in Sydney and Melbourne. Far more commonly, however, the cost of congestion is used as an argument for building more transport infrastructure – usually roads. For instance the President of the Royal Automobile Club of Victoria (RACV) in support of a program of continued motorway building, said, 'The avoidable costs of congestion are predicted to rise to a staggering $20 billion across Australia by 2020 and it is business and the community who will end up paying the bill' (Isaac, 2010).

The staggering amount of alleged congestion costs is primarily derived by summing the small time savings that would be made by individuals if their travel by road were carried out at 'free flow' speeds. That this might be an implausible storyline is not often discussed in the transport planning literature (but see Metz, 2008: 30; Low and Odgers, 2012). How much would it cost financially to build a road system in urban areas that would guarantee 'free flow'? What would be

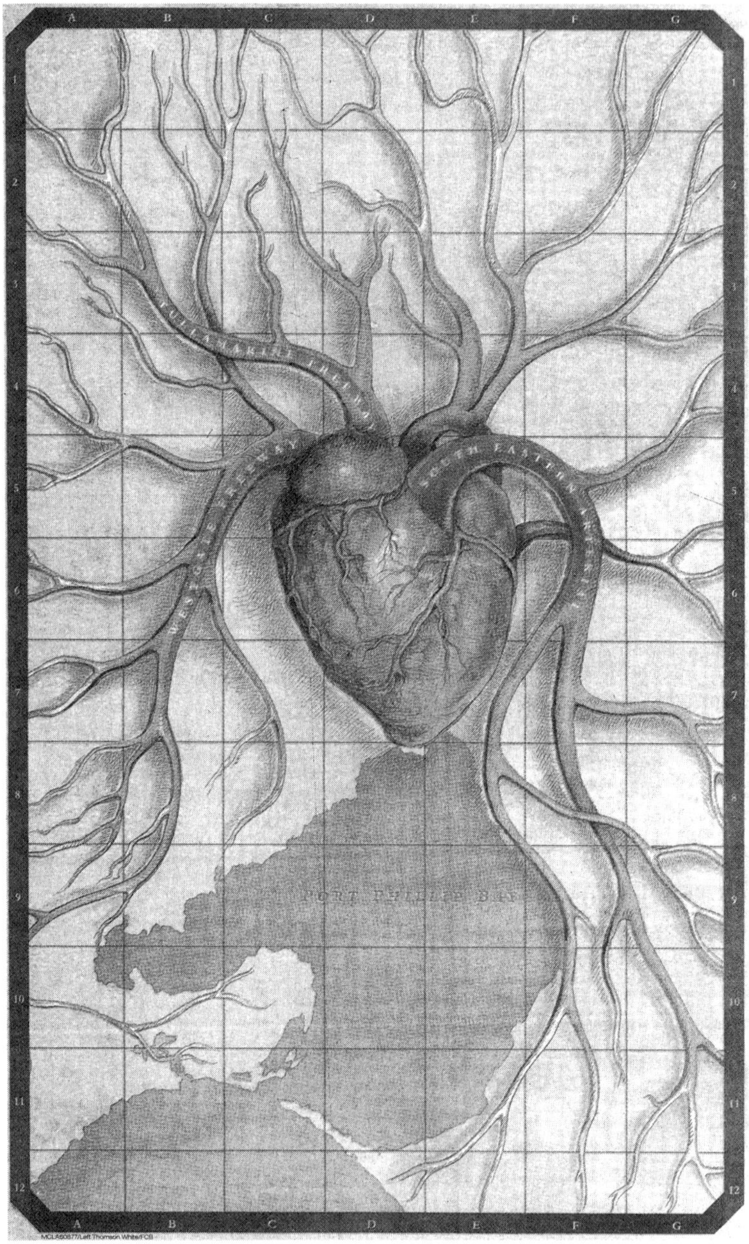

Figure 3.1 Melbourne – An advertisement promoting the City Link Motorway

Source: Image from an advertisement for City Link which appeared in *The Age* entitled 'Life Will Certainly Improve Once We Have a Triple Bypass'.

the environmental cost of such a free flow system? Do individuals really find productive ways of using small time savings (usually a few minutes)? Wouldn't individuals use a free flow system to travel further rather than save time in travel? The answers to such questions tend to suggest that inflated calculations of time savings are actually nothing more than post hoc justifications of bundles of favoured infrastructure projects.

Conclusion

The theory of path dependence provides a challenge to the apparent functional rationality of patterns of urban systems, institutions, organizations and discourses. Pierson (2000: 264) observes that, 'Arguments about increasing returns ... suggest the large dangers in any assumption that an institution arose because it serves some particularly useful purpose ... many alternatives to the outcome in question might have been possible, and a dynamic of increasing returns may have locked in a particular option even though it originated by accident, or the factors that gave it an original advantage may have long since passed away'. By institution Pierson makes clear that he means a congeries of policy settings, discourses and organizations. Transport policy that assumes the functionality of a major bias of expenditure towards improving and maintaining roads for private vehicles is surely a case in point. The institution of roads bias throughout the develop world did not occur by accident but its rational basis is now very much open to question.

The concept of path dependence gives a name and a greater sense of reality to the feeling that sometimes finds expression in various ways: 'there is no alternative'. For instance, with considerable candour the Premier of the State of Victoria (from 1972 to 1990) in an interview by Stone said of transport policy under his administration: "We were tempered by pragmatism – a bit like Lady Macbeth – [sometimes you are] so far up the track that you can't turn back" (Stone, 2009: 426).

If path dependence is suspected, it does not call into question the competence or probity of individual actors in transport agencies. On the contrary their very competence and efficiency in what they do that may be the cause of path dependence; and it is extremely important that such competence is given due recognition. What is necessary is to reframe the problem and redirect the competence. The world of transport policy is populated to a large extent by engineers, and engineers can be very good at what Adams, an engineer from Stanford University, called 'conceptual blockbusting' (Adams, 1974). One of the conceptual blocks Adams writes about is cultural tradition: 'The problem arises when individuals become so universally in favor of tradition that they cannot see the need for and desirability of change in specific areas'. In the case of transport policy there is tendency for actors to fall into entrenched positions: for instance pro-roads, pro-public transport, or anti-mobility, when stepping back and reframing the problem might reveal some way forward that could attract broad political support. But this reframing has to be real, and not a tokenistic gesture to cover tradition in the clothes of innovation and change. As Adams writes, ‘It is often difficult to see a problem from the viewpoint of all the interests and parties involved. However, consideration of such viewpoints not only leads to a “better” solution to the problem, in that it pleases more interests and individuals, but it is also extremely helpful in conceptualizing’ (ibid.: 22).

Path dependence also provides an explanation of the substantial observed differences amongst cities and their transport systems which appear to have developed within similar contextual conditions, sometimes within the same country. For instance, path dependence can add a further dimension to the explanation of the vastly different responses to neo-liberalism amongst the nations (Torfing, 2001).

Path dependence is not a deterministic explanation of local history. It leaves plenty of space for actors to influence events, and for change to occur. But the windows of opportunity for action occur infrequently, and what has to explained is the long periods of stability and continuity of policy. Pierson (2000: 265) comments: ‘Path dependent analyses need not imply that a particular alternative is permanently locked in following the move on to a self-reinforcing path. Identifying self-reinforcing processes helps us understand why organizational and institutional practices are often persistent – and this is crucial because these continuities are a striking feature of the social world’. North (1990: 98–99) cited by Pierson makes the point well: ‘At every step along the way there are choices – political and economic – that provide real alternatives. Path dependence is a way to narrow the choice set, and link decision making through time’.

Finally path dependence provides a coherent framework for research in which a number of other concepts can be enfolded. Even though path dependence does not provide a predictive tool it does provide a reflective one which may in time encourage decision makers to be more critical of existing policy settings and more inclined to investigate from time to time whether they might be in need of change. Pierson argues that ‘recognizing the existence of path dependence may not help policy makers much if they do not know how to identify it ex ante. But this objection loses its force if our purpose is instead to understand – perhaps ex post – why aspects of societies move in particular directions and the consequences of such movements’ (Pierson, 2000: 256). Path dependence opens up an extensive academic field for research in public policy subjecting a variety of policy domains to historical analysis.

In the following chapters we report studies of transport institutions in Australian cities investigating whether the evidence is such that path dependence could stand up as an explanation for what has occurred.

Chapter 4
Transport Planning in Australia[1]

Our case study of path dependence is Australian, and in this chapter we outline the Australian context of transport planning in order to set it in international context. Australian transport planning is influenced by both American and European thinking, but Australia has its own unique history, geography and institutional landscape which itself acts powerfully to influence events. In what follows we first consider the Australian context, starting with its basic urban geography marked above all by distance – both from the rest of the world and between metropolitan cities. Then we turn in more detail to how transport systems are provided – the institutional landscape of transport planning. Finally we identify some of the critical junctures that have shaped Australian transport planning, and we question whether at the present time Australia may be in the midst of one such juncture.

Australia in Context

In Australia there are five large metropolitan, port cities around the coast ranging in population size from 1.1 million (Adelaide) to 4.2 million (Sydney). The main urban centres (including large rural towns) comprise about 60% of the population.[2] The mainland state capitals are surrounded, at a distance of up to 150 kilometres, by smaller rural or coastal towns. The capitals are separated from each other by huge areas of sparsely populated agricultural land and in the case of Perth by the Nullarbor desert. Distances between cities are long. From Perth to Adelaide is 2,725 kilometres (and to Sydney 4,110 kilometres) making Perth among the most geographically isolated cities in the world. From Adelaide to Melbourne is 728 kilometres, Melbourne to Sydney 900 kilometres, Melbourne to Canberra 647 kilometres, and Sydney to Brisbane 1,010 kilometres.

Australia is both like and unlike European urban societies and both like and unlike American urban societies. Like European cities, Australian cities have thriving urban cores which are now sought after as housing locations. There has never been anything resembling the 'white flight' that characterized American cities. Australian cities have well developed public transport systems. Melbourne boasts the world's largest operating tramway system. Both Melbourne and

1 This chapter has been co-authored with Bill Russell and draws heavily on his empirical research and experiences.

2 Population figures taken from Australian Bureau of Statistics estimates for 2005 – http://www.ausstats.abs.gov.au/ausstats/subscriber.nsf/0/80A49E8C.

Sydney have extensive suburban railway systems albeit in need of modernization, extension and better management. Brisbane has extensive new bus rapid transit with underground bus tunnels and stations. Adelaide, long bus-dependent, is now investing in suburban rail electrification and tram extensions. Perth has startled observers by reversing decades of freeway obsession with some remarkably cost-effective and popular new suburban railways.

Despite these public transport strengths, Australian cities have sprawling suburbs equipped with, in American parlance, 'big box' retail, 'strip shopping malls' and vast car-oriented shopping centres. While Australian cities have taken steps to preserve buildings and precincts of 'heritage' significance in the older central areas, CBDs are dominated by tall office blocks and, increasingly, apartment blocks. Transport planning for cities in the latter part of the twentieth century has been dominated by the 1950s American model: the pursuit of free and fast movement by personal transport. Australia has not yet had its SACTRA moment and 'predict and provide' remains the mantra of State treasury departments when approving infrastructure projects. Major new urban motorways are still being planned and constructed, despite several tollways disappointing their promoters with low traffic numbers. Freeway obsession is however beginning to be questioned once again at both State and federal levels, but, as we will argue in subsequent chapters, the inertia of the past carries forward into the twenty-first century.

Australia is a post-colonial society inheriting British nineteenth- and twentieth-century democratic political institutions adapted for the colonies. The Australian 'State' is an institution created at Australian federation in 1901 from the previous group of self-governing colonies. The six States, together with two mainland 'Territories' (The Northern Territory and the Australian Capital Territory) somewhat artificially divide up the whole surface of the continent. With the exception of Queensland they have bicameral Parliaments elected by a variety of different voting systems.

In 1901 the States agreed among themselves, and with the colonial power, to form a federation with a national, federal government called the Commonwealth Government (not to be confused with the British Commonwealth of Nations). Elections for the Australian federal Commonwealth Government are held every three years. It is widely agreed by Australian political analysts that the federal government has steadily enlarged its powers relative to the States with each successive administration. One of the most important federal enlargements was the acquisition the power to levy income tax at the beginning of World War II, following a referendum to amend the constitution.[3]

Elected local governments were instituted by the colonies and therefore pre-date the States, but then as now local governments were the creatures of and subordinate to a higher power, then the British colonies, now the State governments. As in the British constitution, the central power (in this case the

3 Prior to WW II the States levied income tax, but the Commonwealth persuaded the States to cease levying income tax for the duration of the war (following the Commonwealth's acquisition of that power), and State income tax was not subsequently resumed.

State government) has the power to dismiss any locally elected Council, or all of them at once within its jurisdiction, as happened in the State of Victoria in 1993. Australian local governments have limited functions by international standards, and are generally not public transport providers.

What is missing, remarkably, in most Australian metropolitan cities is a *metropolitan* tier of government. In this they are more like American than European cities of similar size. While Sydney, Melbourne, Adelaide and Perth are large metropolitan cities by European and American standards, this fact does not have a place in the governmental system. Only in Brisbane is there an elected City Council (with a Mayor) covering a large part of the metropolis. The local government of metropolitan Melbourne is divided amongst 31 local councils, Sydney amongst 37 local councils and Perth among 32. The city councils of Melbourne, Sydney and Perth cover only the small central areas of the city (and in Perth the State government presided over further fragmentation of the City by creating two new local councils and other boundary changes). Unlike American cities, however, the seat of State government is in the heart of the metropolis. This means that State governments also function as *de facto* metropolitan governments, albeit politically influenced by vast non-urban hinterlands with considerable political clout. Brisbane is only a partial exception because the State of Queensland still wields much control over infrastructure development and funding.

Elected State Parliaments, when they were instituted, were nominally placed in control of the bureaucratic apparatus, but in practice they exerted little actual control. This apparatus, which was already in place under colonial rule, was quite accustomed to conducting its own affairs and getting on with the task of developing the social, technical and physical infrastructure necessary for the functioning of cities, towns and markets. This model of governance, as we will demonstrate in detail in Chapter 5, organized public service agencies in vertical bureaucratic 'silos' with very little horizontal communication among them, let alone co-ordination of their activities and responsibilities. Perhaps this is simply characteristic of 'bureaucracy' itself as an ideal type. Whilst there have been serious attempts from time to time by State governments to change this system, particularly with the aim of giving increased power to elected State Ministers and Cabinets to rule and steer the bureaucracy (see Gleeson and Low, 2000, chapters 4 and 5), the bureaucratic silos remain present to some extent in the culture of Australian governance. This silo effect is especially important in transport where the road planning agencies, and each separate form of public transport agency (trains, light rail or trams, buses), have historically had their own organizations with different degrees of power and different ways of viewing the world (described in Chapters 5 and 6). Land use planning, likewise, normally exists in a silo of its own. There have been efforts to integrate the management of public transport, with the exception of the Perth case, these have had only limited success.

We now turn to the finer detail of the institutional landscape of Australian transport planning. We combine an examination of functions at different levels of government with the financial resources available to different levels.

The Institutional Landscape of Transport Planning

Russell's historical examination of the role of public transport planning in three federal systems (the USA, Germany and Australia) provides a useful vantage point from which to view Australia's institutional landscape (Russell and White, 2009). Russell compares the situation in Australia with the aggressive support for road building in the USA coupled with the systematic undermining by General Motors and some city governments of the development of public transport. In Germany he identifies the critical role of the Third Reich in supporting the car industry and initiating motorway construction. Even though public transport flourished in the Bundesrepublik after the war, the historical role of the Nazis in promoting the car and the road as the model for future transport needs to be recovered.

In Australia Russell identifies three waves of involvement at Commonwealth (federal) level, all supported by Labor Governments: the Whitlam government's Public Transport Improvement program, the Building Better Cities program of the Hawke-Keating governments, and the Rudd government's Infrastructure Australia initiatives. All have been driven by a federal concern for improving city life. However, State governments have continued to exercise primary responsibility for public transport and to provide the bulk of funding for it. The Commonwealth government has played an important but ancillary role in stimulating innovation and providing impetus and funding for projects that might not otherwise have proceeded. In summary Russell observes:

> In short, we can say that Commonwealth intervention in support of urban public transport has been important, innovative, but discontinuous and unreliable. By contrast Commonwealth involvement in providing sustained support for urban freeways has been bipartisan and continuous. Such policies, combined with laissez faire urban development policies among the States, have helped to shape the high-carbon lifestyle and ultimately unsustainable pattern of development in Australia's larger cities. (Russell, 2010: 5)

We now examine the role of each tier of Australian government in transport planning. We focus mainly on the federal level, leaving a more detailed examination of path dependence situated at State level to subsequent chapters.

The Federal Tier

The Federal Constitution gives the Commonwealth of Australia power, under Section 51, to acquire the railways of any State and to build interstate railways (always with consent of the State). The Commonwealth exercised these powers to build interstate railways, but not urban passenger railways. Because the main function of the road and rail systems at national level is to link agricultural areas and remote mining centres with the metropolitan ports, the federal transport portfolio has traditionally had a non-urban bias. With cities so distant from one another, national expenditure on non-urban roads necessarily still takes a large share of the roads budget. A continuing theme of federal interest in urban transport has been road, and occasionally rail, access to the ports.

The Commonwealth's involvement in land transport was extended to roads in 1922, and for the next fifty years progressively greater amounts were provided as grants to the States. In 1924, William Calder, the Chairman of the Country Roads Board of Victoria visited the USA and Europe to study roads development. His report was quoted at length in the Commonwealth Parliament during debate on the Federal Aid Roads Bill (mirroring the US Federal Aid Roads Acts of 1916 and 1921). Calder recommended the hypothecation of petrol tax for road construction and federal intervention in highway financing. Calder had also in 1924 initiated the Victorian Highways and Vehicles Act which provided for the declaration of state highways and their joint financing by State and local governments. Calder's report was the medium by which the principles and administrative arrangements for road financing were passed from the USA to Australia.

Wartime mobilization resulted in the acceleration of Commonwealth involvements in strategic transport projects, and through the 1950s and 1960s, strategic roads such as the Beef Roads program (for agricultural purposes) were developed. However, despite its growing role in road financing, the Commonwealth at this time had no involvement in urban public transport.

During the 1960s in America freeway building began to be questioned. A turning point of a sort for public transport came in 1962 when the incoming President John F. Kennedy, called for 'the needed expansion of public mass transportation' and 'balanced transportation' (Norton, undated).[4] In 1970–71 in Australia, total Commonwealth expenditure on roads amounted to AU$252 million, increasing each year by about 10% to reach AU$608 million by 1979–80. However building roads also began to be questioned in Australia, assisted by the first oil shock of 1973 and community protests against urban freeway building plans. Several of these plans were prepared under the tutelage of US freeway consultants: Wilbur Smith in the Melbourne Transportation Study (Metropolitan Transportation Committee, 1969) and the Sydney Area Transportation Study (NSW Department of Transport, 1974), and Vorhees and Associates from Virginia on the Perth Regional Transport Study (Metropolitan Region Planning Authority, 1970). In March 1973 the Premier of Victoria, Rupert Hamer announced the cancellation of 150 miles of the 307 miles of freeways that had been proposed by the Metropolitan Transportation Committee following the Wilbur Smith study. Hamer later stated that freeways were 'most intrusive and damaging to the environment in the inner city areas. We were prepared to put up, in those areas, with some congestion and delays rather than disrupt communities' (cited in Anderson, 1994: 244).

4 The storyline of Balanced Transport is long lived and pervasive. In Western Australia, for example, it is the main objective found in the Department of Transport's 1995 Metropolitan Transport Strategy (Department of Transport, 1995).

By the end of the 1970s the proportion of Commonwealth road spending that was devoted to urban freeway construction had fallen to 25% (AU$127 million of a total of AU$535 million). However, following the dismissal of the Whitlam (Commonwealth) government, spending on rural roads grew, devoted largely to expenditure on the National Highway System under Whitlam's *National Roads Act*, 1974. At the start of this period the Whitlam government began to explore the possibility of intervening in urban public transport. Under Whitlam, the Commonwealth established the Urban Public Transport Program through which the Commonwealth provided grants over a five year period under the States Grants (Urban Public Transport) Act of 1974 specifically for public transport in the metropolitan cities. The lion's share of funding under the program went to Sydney and Melbourne for urban rail projects, including a new East Hills to Glenfield line in Sydney and upgrading of the Ringwood line in Melbourne. The first phase of the program ran from 1973 to 1978 during which the Commonwealth provided AU$188 million. The second phase of the program was meant to last five years from 1978–83 but was prematurely terminated in 1981–82 following a cost cutting exercise by the Fraser government.

In this period the Commonwealth expenditure on urban public transport, although significant, was small compared with the continuing expenditure on roads. The Commonwealth dominated the freeway component of road funding, and spending on urban arterials amounted to AU$1.4 billion over the decade 1970–1980. State expenditure on urban arterials amounted in aggregate to AU$1 billion with annual expenditure rising every year from AU$61 million in 1970–71 to AU$197 million in 1979–80.

The level of Commonwealth expenditure on roads almost doubled under the governments of Hawke and Keating: from AU$862 million in 1982–83 to AU$1,601.7 million in 1996 reaching a peak in 1992. The National Highway System accounted for 40.7% of all roads spending for the period 1982 to 1992 but thereafter jumped to 50%. A number of roads programs were pursued in the 1980s. In the years 1983 to 1991 the Commonwealth government spent over AU$4 billion on the National Highway System with greatest emphasis on interstate highways linking Melbourne, Sydney, Brisbane and Cairns (in north Queensland). Importantly from the end of 1992–93 the funding for arterial roads was untied – that is not directed at specific Commonwealth programs, and after 1997–98 determined purely on a *per capita* basis.

Whilst road funding continued to grow, the Hawke-Keating government also resumed funding for urban public transport. Under the Australian Bicentennial Road Development Trust Fund Act 1982 States were given the option of using up to 25% of funds allocated for urban arterial roads for urban public transport capital projects. Acknowledging that the Commonwealth had been funding public transport since 1982, the Department of Transport increased its commitment to public transport and established a separate program in 1991–92 (Commonwealth Department of Transport Annual Report 1990–91: 107). Spending on urban public transport projects 1990–1993 amounted to AU$191.78 million. Two examples from Sydney were the 'Y' Link a 1.6 km rail line between Merrylands and Harris Park stations, allowing trains to travel directly between the southern and western lines, and the dedicated bus lane on Sunnyholt Road at Blacktown. These projects were part of the Building Better Cities program (Building Better Cities, 1993).

The Special Premiers' Conference also initiated the Building Better Cities (BBC) program with the purpose of 'improvement of urban consolidation and the quality of urban life' (*Building Better Cities*, Auditor General of Victoria, 1996). BBC incorporated a series of projects grouped as 26 'Area Strategies around the country. Of these, 11 projects involved public transport components. BBC project allocations amounted to AU$816.4 million between 1991–92 and 1995–96. A project funded in Melbourne was the City Circle Tram Loop, which provided free travel around the CBD. Data collected by the Public Transport Corporation suggested that over 3 million passengers used the service in 1996. The largest category of users was people employed in the CBD (59%) followed by tourists (35%). A similar project in Perth provided the CBD with a free central area transit system (CATS) which continues to run today.

Under the Howard Government that succeeded that of Keating in 1996 the level of road funding totalled AU$1,616 million (2010 prices) and remained relatively constant, reflecting a continuation of the level of funding which existed under the Keating government in the early 1990s through to the end of the decade. The Bureau of Infrastructure, Transport and Regional Economics (BITRE) records Commonwealth road expenditure (in 2004–05 prices) as AU$2,477 million in 1985–86, rising to a maximum of AU$3,043.5 million in 1992–93 and dropping to around AU$2 billion per year thereafter (AU$ 1,933.3 billion in 2003–04).

The Roads of National Importance program (RONI) was implemented in 1996–97 in recognition that many roads not included in the National Highway System provided a range of social and economic benefits which made them worthy of federal funding. Such roads included outer suburban roads such as the Scoresby Freeway, the Pakenham Bypass, and the Geelong Bypass in Victoria (though federal funding was withdrawn from the Scoresby Freeway when the State Government designated it a tolled public-private partnership project). The political nature of RONI quickly emerged. Although funding for major road projects is often announced in the lead up to elections, the RONI project gave politicians the opportunity to fund smaller projects not all of which seemed to meet the funding criteria.

Other road funding programs included the Roads to Recovery program (R2R), a four year program with the promise of AU$1.2 billion for local roads by June 2005, and the Accident Black Spot program designed to improve the design of roads and intersections to decrease the risk of fatalities and serious road injuries. Funding of AU$36 million was provided in 1996, rising to AU$45 million by 2003–04. Most of this money went to improving road access and safety for car users (sometimes at the expense of other road users where roundabouts were the favoured solution to resolve car-car crashes at intersections in suburban streets). It has been possible for local councils to spend R2R funds on bicycle infrastructure and one local council in Perth chose to spend 100% of the grant in this way.

In 2005 the Federal Government enacted new legislation to govern its involvement with land transport: the AusLink (National Land Transport) Act 2005. AusLink applies to the major road and rail links which are assumed to run in 'corridors' such as those between each state capital. Though it includes a few interstate rail freight projects, the AusLink program is overwhelmingly comprised of roads projects devised by State government road planning and construction agencies and offered for funding to the Commonwealth Government.

The first phase of AusLink (AusLink 1) provided funding of AU$9.3 billion over five years (2005–06 to 2009–10) for national projects on the AusLink national network. Of that only AU$120 million was allocated to national rail projects, and that was for the upgrading of bridges to bear heavy vehicles.

The next phase of AusLink was designed to cover the five years from 2009–10 to 2014–15. Bids from state governments for this funding were made in July 2007, and in general greatly exceeded the funds available. For example Queensland sought AU$30 billion under AusLink 2 and Victoria AU$11.5 billion. There was considerable variation amongst the States in how AusLink 1 money was spent. In Queensland, it was used to fund urban freeways, in New South Wales it was committed to rural freeways with some funding for rail freight development, and in Victoria it was earmarked for urban and rural freeways with a small allocation for rail, mainly for port access.

In 2007 the Howard government was defeated and the federal role in Australian land transport was significantly changed with the election of the Rudd government. Four key elements were important in distinguishing the Rudd government's approach from that of its predecessors.

1. The government's enthusiastic adoption of the 'infrastructure gap' storyline that had been promoted from 2006 onward both by local government peak bodies and by the lobby group for private infrastructure providers;
2. The creation of significant changes to machinery of government to implement a program to fill 'the gap' including a federal Minister for Infrastructure, a new agency: Infrastructure Australia, and the renaming of the Commonwealth Department of Transport and Regional Development (DOTARS) and the Bureau of Transport and Regional Economics to include the word 'infrastructure' in each case. A new policy focus was developed in which a national approach to transport policy would be created through COAG in which all State transport ministers were assigned roles to develop aspects of national policy.
3. The creation of a Building Australia Fund, and the provision of some AU$20 billion through the 2008 federal budget for infrastructure projects.
4. The signalling of a return of Commonwealth concern for the cities (a reference to previous Labor government initiatives under the Whitlam [1972–75] and Hawke-Keating [1982–96] governments.

In April 2009 following Council of Australian Government deliberations, Minister of Infrastructure Albanese announced a AU$32 billion program of infrastructure funding mostly for new roads. Subsequently, however, under the heading 'Nation building infrastructure for productivity and prosperity' a more evenly balanced contribution to new expenditure in the 2009–10 federal budget was presented with AU$4.623 billion going to rail projects, AU$3.441 to road projects and AU$389 million to port infrastructure (Commonwealth Budget Infrastructure Overview, 2011).

These projects were launched in the context of the national economic stimulus as the government responded to the global financial crisis. For the first time public transport and rail based projects came to the foreground of Commonwealth attention, with a new Regional Rail Link in Melbourne at AU$3.4 billion being the largest project funded. This seemed to herald a new critical juncture, similar to the generational victory achieved by the roads lobby in 1942, discussed further below.

The State Tier

The States receive about half their revenue from the federal government through grants. Thus federal policies regarding competition policy and road funding are very influential on State policy outcomes. The other half of their expended revenue comes from State taxes on land, payroll, financial transactions, motor vehicle registration, gambling (Commonwealth Treasury 1999) and from revenue raised through the activities of government owned enterprises.

Apart from the Auslink corridors, State governments are responsible for the production of all new roads. State governments are also responsible for provision of public transport and maintenance of State highways and lesser roads. Finance for the development of this transport infrastructure has been obtained through a number of mechanisms ranging from:

- allocation from the state budget
- direct borrowing and/or debt financing
- franchising agreements which include arrangements for infrastructure development
- public/private partnerships (Build-Own-Operate, or Build-Own-Operate-Transfer projects)

As noted above, the states have all traditionally managed the provision of transport services within separate bureaucratic 'silos' as part of the Australian public administration system. This has led to the separation of functions relating to transport: aviation, roads, and public transport (buses, trains and light rail – trams). Due to the availability of funding, the emphasis on road development over a long period, and the long term employment tenure of public servants, there exist embedded and powerful groups of road engineers in the public service.

These engineers actively engage in planning and developing projects, and tend to view transport solutions in terms of road building projects (for the history of the ideology and institutions of road building see Davison with Yelland, 2004).

The delivery of public transport is also generally in the hands of the State governments, or their contractors. But, again with the partial exception of Perth in Western Australia, States have not been successful in integrating the bureaucratic silos responsible for different modes of public transport. Moreover, since operational funding for the cities' public transport systems comes out of State budgets, there has been a continual strong concern about the 'public transport deficit' (a storyline discussed in Chapter 9) and a corresponding emphasis on cost saving. Expenditure on public transport is depicted as a cost, whereas expenditure on roads is depicted as 'investment' – two very different and compelling stories. The primary work of roads engineers has been to build new infrastructure whereas, again with the recent exception of Perth, the primary work of railway specialists is to operate the railway system at least cost. The operational costs of private transport on roads are of course very high but they are individualized and hidden from public view.

During the 1980s and 1990s the silo system was reinforced by competition policy advocating micro-economic reforms, in particular the break-up of vertically integrated monopolies. This was achieved through the separation of policy and service delivery (the separation of 'steering' from 'rowing'). All States have implemented this sort of reform in public transport service provision. Interestingly in almost every state the roads corporation has survived relatively unscathed. State roads authorities have used private roads contractors for many years. Obviously because of the ongoing nature of maintenance there is a large support industry for roads in the private sector (gravel and concrete manufacturers, asphalt firms etc). This nexus of interests in road building has generally been supported by urban land developers and private engineering consultants. Thus the impact of microeconomic reforms on roads authorities (responsible for planning and maintaining State road systems) may have been reduced, being seen more as the continuation of a trend. It also means that a significant private sector nexus (within and inside government) for the continued development of roads exists which does not exist in the public transport sector.

Over time the approach to transport policy by the States has varied with different governments, and under the impact of different ideologies in fashion. These shifts have had consequences for travel behaviour in the main metropolitan cities, and for the present bureaucratic structures for infrastructure policy. In the 1970s and early '80s the government of New South Wales placed a strong emphasis on railway modernization, although the network remained poorly designed and a multi-billion dollar project to 'untangle spaghetti' was later required. In Queensland the recent establishment of a transport authority for the South East Queensland metropolitan region (centring on Brisbane) has the potential to create an integrated service structure, overcoming long established divisions between bus and rail operations, fragmented structurally with State government owning the rail system and Brisbane City Council the bus system.

The Dunstan Government of South Australia in the 1970s stopped further motorway construction, extended suburban railways, took over the private bus operations, and integrated fares and timetables across the public transport system. But in the 1980s these progressive policies were abandoned, motorway construction was resumed, and public transport was limited to buses. Adelaide became the last state capital to commence electrification of its suburban railway system.

Western Australia has moved furthest to integrate transport policy and attempts have been made to centralize the control of public transport which remains a public sector function. Perth has seen major extensions to its railway system both to the north and south of the city.

Critical Junctures in Australian Transport Policy

It is evident from the above discussion that, whilst funding for public transport has occasionally appeared on the government agenda, the dominant transport paradigm has been to accommodate motor vehicles by building roads. Path dependence theory holds that such a path was initiated at a 'critical juncture' in public policy making. What then was the critical juncture that produced the policy path dominated by roads and cars, and when did it occur? A critical juncture as Mahoney (2000) argues is not a single event but a cluster of events over a relatively short period of time in which change occurs to established policy patterns, a period in which a choice is made amongst alternative paths from which it becomes increasingly difficult to retreat. In transport it is, as Imran (2010: 66) describes, 'a period in which certain social forces combine to change the pre-established pattern of transport in the country'.

In Australia this period occurred between about 1943 and 1956. Davison (2004, Ch.5) records how, during the second world war and in the immediate post war years in Australia, the public aspiration to car ownership grew, fuelled by an active lobby of manufacturers and automobile enthusiasts. But the task of building new roads to accommodate high speed private cars was in the hands of government, and governments (federal and state) at that time were focused on the task of mass public transport in the cities.

The year 1943 saw two events that in Davison's view, 'signalled an important shift in the political landscape' (Davison with Yelland, 2004: 116). The first was the rise of Robert Menzies to head a young group of Liberal activists (the National Service Group) 'determined to … promote an alternative vision of the post-war social and political order' (ibid.: 117). The second was the revival of the Australian Automobile Association (AAA) by Sir Thomas Nettleton, president of the RACV, to bring together the interests of the State motoring associations and act as an 'aggressive national lobbying organisation' (ibid.). The AAA subsequently sought to redirect the fiscal and regulatory powers of the national government 'towards the democratic goal of mass motorisation'. Building the road infrastructure to support this goal, however, required strong public sector intervention – already instituted during the war – which the AAA also advocated: planning, coordination, organization and national development. Davison notes that the goal of the AAA was to create a national agenda for motorisation. This agenda included five prime goals:

> Removing restrictions on the supply of petrol and imported motor vehicles; establishing the car as the dominant form of overland transport for both goods and people; directing revenues derived from taxes and charges on motorists to road construction; standardising and simplifying regulations; promoting road safety, and research and planning on road issues. (Davison with Yelland, 2004: 121)

In his speech to launch his successful election campaign of 1949 Robert Menzies, leader of the rebuilt and revitalized Liberal Party, linked the stabilizing virtues of 'home' (sanity, sobriety, continuity and health), with the dynamic virtues of the motor car (liberty, movement and change). Menzies promised to end petrol rationing and urged voters to 'empty out the socialists and fill the bowsers' (Davison with Yelland, 2004: 112). Petrol rationing was lifted three months later (February 1950). Menzies hailed it as 'an earnest of the new freedom … to which the Commonwealth has pledged its endeavours' (Davison with Yelland ibid.. citing Broomham, 1996: 96). Menzies thus inscribed the core of the agenda of the AAA in the national political program: establishing the car as the dominant form of overland transport for both goods and people.

This agenda was not without serious opposition. State governments and their bureaucratic actors were focused on providing the crowded public transport services. Both the public transport unions and the powerful State bureaucrats in charge of public transport modes were influential within the Labor Party. In the early 1950s in Victoria former railway workers held four of fourteen Cabinet positions in the State. As Davison observes:

> The contest between private and public transport was also a class question. Public transport was patronised, as well as owned, by the people. Defending the train and tram against the car meant upholding the interests of the majority who could not yet afford cars against the vociferous minority who could. (ibid.: 125)

Nevertheless the rapid growth of motor car ownership began to exert its own effect. The year 1945 was the last year the railways in Victoria operated at a profit. Patronage of the rail and tramways declined and what was formerly a majority mode of urban travel became a minority mode.

In Victoria the year 1956 can be described as the end of the period of critical juncture when the dominant path had become entrenched. The Victorian Parliament under the leadership of the newly elected Liberal Premier, Henry Bolte, passed three Acts which had a major impact on the institutional structure of the transport portfolio: The Melbourne and Metropolitan Board of Works Act, The Road Traffic

Act, and the 1956 amendments to the Country Roads Act. The first united the functions of town planning and highway design and construction in one agency, the Board of Works. The second created a body that would determine uniformity of traffic management across the State regulations, and the third amended the Country Roads Act to empower the Country Roads Board to acquire land for the construction of by-passes (which later became known as freeways), to control access to these roads, and to build and maintain them. These acts by the Bolte government marked the beginning of the institutional empowerment of the roads and private car agenda and the subsequent steps towards consolidation of bureaucratic functions in the roads domain around what later became the Roads Corporation of Victoria (VicRoads). Similar processes occurred in other States, though as we show in the next chapter, the path of road-car institutional dominance took somewhat different forms in the three cities we examine: Melbourne, Sydney and Perth.

The establishment of Infrastructure Australia by the Rudd Government in 2008, and subsequent decisions to allocate major funding to urban rail projects, may be the next critical juncture in Australia's public transport development. As argued earlier, Commonwealth intervention has repeatedly supported innovation in transport policy, and in successive waves has supported new urban public transport initiatives. Is it possible that the new rationality of national audit, established through Infrastructure Australia and embraced by the Commonwealth Government, may be the critical juncture at which Australia disengages from the Menzies vision of a car-dependent nation? In the interests of liveable cities and a sustainable future, this is something to be hoped for.

Conclusion

Australia's urban transport policy development is at least in part governed by a historical desire to provide equivalent services across a geographically massive area, with low population density to pay for it. Despite the redistribution of wealth in this way, there is still a large discrepancy between the standard of services (particularly health, public transport, energy and telecommunications) provided to rural people and those in urban areas. The functional domain which most successfully reflects this redistribution of wealth is roads.

Roads and cars are deeply embedded in the Australian economy as well as in the Australian popular culture, leading to claims such as that made in the *Financial Review* newspaper recently that a carbon tax would add AU$412 to the price of a typical Australian car and 'threaten the viability of the local manufacturers'.[5] Similarly the construction company Boral has an entire division structured around the provision of bitumen. Concrete companies also depend significantly on road infrastructure construction. The conservative Commonwealth Government (up to

5 Financial Review 13 May 2011 http:afr.com/p/national/carbon_tax_will_wreck_car_industry_3KKw8VxZq4YnDqUZzPo10J (accessed 30 May 2011)

24 November 2007) was opposed to policies which might change the structure of Australia's economy in any fundamental way, including its adaptation to the exigencies of climate change.

Australian transport policy has too often been compartmentalized geographically, institutionally and functionally. Cities are separated by huge distances, and notwithstanding the Commonwealth government's efforts to overcome the tyranny of distance, they remain seriously time-segregated in land transport. While a healthy 'pressure group' politics surrounding the delivery of metropolitan transport services and land use planning is focused at State level, much power to determine outcomes lies at federal level through its command over large scale funding. At federal level the politics is not focused on service delivery, or on urban development, but on the national economy and strategic hand-outs (pork barrelling) at budget and election time. Much of the federal politics of transport takes place behind closed doors in meetings between State and federal ministers at the Council of Australian Governments. At these meetings, transport infrastructure for the next 20 years has been locked into a road-dominant strategy which appears to be completely at odds with any serious strategy to address climate change.

Australia may have reached a new critical juncture. There are major new urban and regional rail projects and light rail projects being implemented across the nation. There are useful bus rapid transit projects and new ways of using buses to provide connections to local communities too long isolated. Infrastructure Australia and the Commonwealth Government have shown that new freeways do not have to be the centrepiece of transport planning and investment. It may be that the era of car dependence, created with such partisan intent by Menzies, is finally losing momentum. Whether this momentum can survive a change of government remains to be seen.

Chapter 5
How Organizations Shape Infrastructure: Roads Organizations

In Chapter 3 we divided the institutional basis of path dependence into two main aspects: the organizational and the discursive. The former is, so to speak, the institutional 'hardware', the rule-bound social structures that enable collective action, the bureaucracies. The latter is the institutional 'software', the ideas and storylines that direct and justify collective action. In this and the next chapter we examine the organizational hardware for transport planning and delivery, how it developed over time in the three cities, and how its development might have contributed to the path dependence of public policy. We map the structural changes to the transport institutions that have occurred and compare them across the three cities. In this chapter we examine the organizational changes to the roads agencies. In the next we examine the changes to the public transport organizations. This 'structural' history sets the scene for the discussion of transport plans and the discourse of transport policy that follows.

Structural Change and the Span of Operational Control

Modern management theory stipulates that effective control of stable technological systems such as road or public transport systems requires a 'mechanistic' (contrasted with an 'organic') structure, with as much as possible of the management task brought under central control (Shafritz, Russell and Borick, 2009: 254). This model embodies four main principles: specialization, unity of direction, centralized authority, and a scalar chain of command (Gibson et al., 2009: 406). The road planning authorities, as we will show, tend to display aspects of these qualities in varying degrees in the three cities. Specialization has been formed around the road planning and management functions, and different aspects of the management task have accumulated around road agencies over time. The specialization of road planning and management is now beginning to be further modified by the need for integrated transport planning, or land use-transport integration. How far the management task can be broadened without losing 'specialization' remains to be seen. The unity of vision of free movement for road vehicles is still very strong, though, again, it has been eroded by the concerns of environmentalism and urbanism. Within roads organizations the chain of command is still intact, reaching up to ministerial level. Yet the need to negotiate with agencies pursuing other functions increasingly limits the degree to which the 'free movement' vision can be pursued. And political intervention can interfere radically with the road agencies' agendas.

Melbourne

Organisational control over road planning and management has been particularly strengthened in the State of Victoria by the continuous consolidation of roads and traffic agencies that formerly performed different functions into one powerful and long standing agency, now called the Roads Corporation of Victoria or VicRoads. The early institutions in the roads sector were designed to perform a single function. The function of the Country Roads Board (CRB) was to build and maintain roads, the Transport Regulation Board (TRB) was responsible for rules governing the use of commercial goods and passenger vehicles, the Motor Registration Board (MRB) licensed drivers and registered private vehicles, and the Traffic Commission (later Road Safety and Traffic Authority – RoSTA) was responsible for road safety measures. Three major changes to this single function framework have occurred, each leading to greater integration of functions.

The roads sector organizations experienced significant change during the past century (Anderson, 1990; Lay, 2003; VicRoads annual reports 1989–2005). The Country Roads Board (CRB) was created in 1913 to improve and maintain main roads throughout the state. Other types of roads were added in subsequent legislation such that by 1958 the CRB was also responsible for developmental roads, State highways, tourist roads, forest roads and stock routes (Victoria, *Country Roads Act, 1958*). Whilst the CRB was in principle responsible for all the main roads throughout the State, in practice most of its work in the early years focused on improving rural roads which were poorly serviced by local councils. Lay (2003: 48) reports that in the early 1950s, 'the CRB was still spending only 7% of its funds on Melbourne roads'. But by 1964–65 the figure had risen to 24%. The limit of CRB intervention in Melbourne was in practice defined by the area covered by the tram network (mostly in central and inner Melbourne) since the legislation of the time prevented roads carrying tramlines from being declared 'main roads'. However, the Country Roads Act of 1948 gave the CRB power to compulsorily acquire land for future road widening, allowing it to expand its work in areas of future metropolitan growth. Lay (ibid.: 49) remarks that, 'It was a wonderfully successful policy in an expanding metropolis', as road alignments and reservations were secured in the outer suburbs to serve future low density urban development.

Within Melbourne, the principal responsibility for roads was divided between local councils and the Melbourne Metropolitan Board of Works (MMBW), a metropolitan authority based on the concept of the London Board of Works which later became the London County Council. Lay reports a personal comment from the Deputy Engineer in Chief of the CRB that, 'this *de facto* policy based on the tram system would leave the MMBW with almost intractable concerns with network deficiencies in the established areas, whilst the CRB could concentrate on the relatively simpler problems of providing for future growth in the undeveloped metropolitan fringe' (Lay, 2003: 49). This function has been extremely persistent up to the present when new outer suburban motorways have continually been built, and more are being planned by the CRB's successor (VicRoads) to serve future outer metropolitan development.

In 1949 the *Town and Country Planning (Metropolitan) Act* empowered the MMBW to produce a 'planning scheme' for metropolitan Melbourne including a plan for the road network. The result was the 1954 Metropolitan Plan (or Planning Scheme). In 1956 the conservative government of Victoria gave the MMBW power to implement its plan for roads, and road planning for a time became divided between two authorities: the MMBW within metropolitan Melbourne, and the CRB for the rest of the State – though there remained some overlap in implementation. MMBW metropolitan planning developed during the 1960s and incorporated the recommendations of a report prepared under the tutelage of Wilbur Smith for a network of freeways at 5 kilometre intervals across the metropolis (Metropolitan Transportation Committee, 1969). The MMBW plan of 1971 (which updated and replaced the 1954 Planning Scheme), which would have had a destructive impact on the local environments of inner and middle Melbourne, was strongly resisted by inner urban communities and local governments, and the State government was forced to cancel the inner core of the proposed freeway system. Subsequently two reports to the Victorian Government recommended that all responsibility for metropolitan main roads be transferred from the MMBW to the CRB (the *Report on the Victorian Land Transport System*, 1971/72, and the *Report on Local Government Finance*, 1972: see Dingle & Rasmussen, 1991). The recommendations of these reports were implemented two years later (by the Victorian *Metropolitan Bridges, Highways and Foreshores Act*, 1974). This was the first step in a process of continuous consolidation of roads planning and management for Melbourne.

Meanwhile, the planning of motorway extensions by the roads authority on its own initiative was continually undertaken. For instance, two years after the Premier had put a stop to new inner city motorway construction, an internal report of the Country Roads Board dated April 1976 shows investigations with quite detailed route plans to connect the F2 freeway in the north of Melbourne with the Eastern Freeway cutting through the inner suburbs. The route plans were marked: 'planning proposal being investigated, FOR DISCUSSION PURPOSES ONLY' (sic) (Country Roads Board, 1976).

The first major change to both the roads and public transport functions occurred in 1983 under a restructure of the whole land transport portfolio. At this time the TRB and RoSTA were amalgamated, combining the functions of road safety and regulation. This new institution, the Road Traffic Authority (RTA) allowed sharing of information and expertise between road accident investigators and staff engaged in designing and enforcing regulatory instruments.

A second change occurred in 1989 when the sector was further consolidated. All road activities – road construction, road safety, driver licensing, vehicle registration and traffic management – all became the responsibility of a single institution, The Roads Corporation of Victoria (VicRoads). Since the 1989 consolidation, VicRoads has retained its multi-purpose form, with some internal realignment of responsibilities. This process of consolidation left VicRoads with a broad span of control over almost the whole road transport sector – including traffic management, responsibility for accommodating road-based public transport, and for bicycle paths, but the question of integrated transport planning inclusive of both private vehicles and public transport moved from academic thought to the political agenda in the twenty-first century. A succession of transport plans was produced that joined these concerns and espoused integrated transport planning (see below).

The third major change towards integration started in 1996 when VicRoads was brought within a Department of Infrastructure (DoI) having responsibility for both transport infrastructure (at the time this meant roads) and land use planning. In this Department, instituted by a neo-liberal government, road engineers from VicRoads dominated metropolitan planning and occupied the senior positions. Ten years later, under a Labor Government, organizational integration faltered with the transport and urban planning functions of the DoI split between a Department of Transport (DoT) and a Department of Planning and Community Development (DPCD). Despite this, at least transport integration was now possible with DoT responsible for both roads and public transport. Ministerial responsibility within the DoT, however, was divided between roads and public transport, each sector reporting to its own Minister. Through both these organizational developments VicRoads retained its integrity, knowledge base and span of control, exerting powerful influence within the Department of Transport (see Astle 2007; Low and Astle, 2009).

The process of 'integration' of transport functions was taken a significant step further with the Transport Integration Act (2009). The Act 'requires all Victorian transport agencies – including the Director of Public Transport, V/Line (the rural rail service authority), and the Linking Melbourne Authority (a statutory authority created to manage the construction of three new motorways) – to work together towards the common goal of an integrated and sustainable transport system' (DoT website www.transport.vic.gov.au accessed 8/10/10).

It is too early to say whether such integration will occur, but evidence that VicRoads is still independently pursuing its own plans to extend Melbourne's motorway system emerged in October 2010 (Lucas, 2010). These plans were described by the VicRoads executive director of roads planning as the outcome of a 'brainstorming exercise' (Lucas, 2010b). The exercise was evidently conducted by VicRoads acting on its own. The proposals include an inner motorway ring and tunnel around the CBD, a connecting North-South motorway through the inner Eastern suburb of Camberwell, and a new outer ring road, as well as other motorway extensions and connections. Such an exercise on the part of a roads planning agency is to be expected. But what has not been seen is a comprehensive analysis of the transport choices for future Melbourne in which the public transport solutions are considered together with and as alternatives to solely road based solutions.

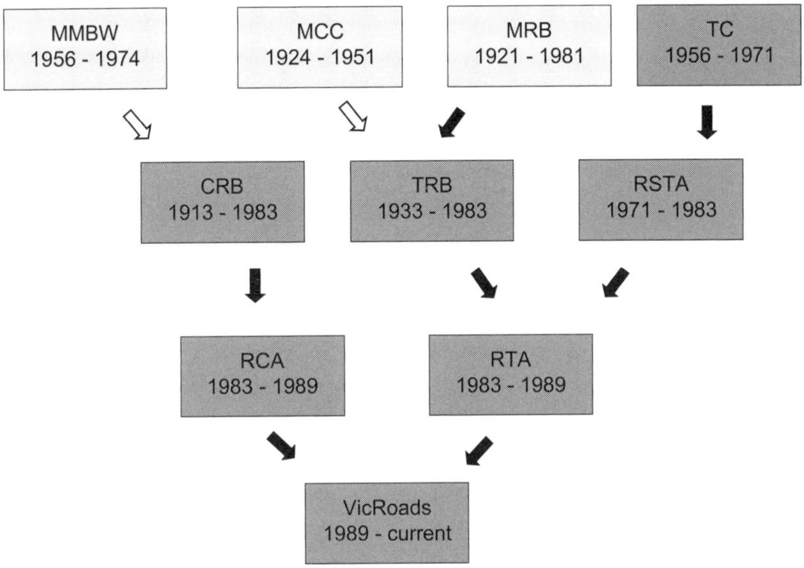

Key: CRB = Country Roads Board; MMBW = Melbourne Metropolitan Board of Works; MCC = Melbourne City Council; MRB = Motor Registration Branch; RSTA = Road Safety and Traffic Authority; RCA = Road Construction Authority; RTA = Road Traffic Authority; TC = Traffic Commision; TRB = Transport Regulation Board; VicRoads = Roads Corporation of Victoria.

Figure 5.1 Melbourne: Structural change in the roads sector

Sydney

As in Melbourne, so also in Sydney (NSW) the organization at the heart of roads planning is a State Government agency: the Department of Main Roads of New South Wales (DMR) and its successor the Roads and Traffic Authority (RTA). Responding in part to Commonwealth initiatives to finance road construction, steps towards creating a central NSW State authority for roads began shortly after the Country Roads Board was set up in Victoria in 1913, but were thwarted by political instability in State politics (Department of Main Roads, 1976: 76). In 1924, with the enactment of the Main Roads Act of that year, the Main Roads Board of New South Wales was instituted. This Act gave the Board, consisting of three members, the power to act as a State roads authority (Rattray and Sinclair, 1978: 18). Consolidation of functions followed and the NSW *Transport (Division of Functions) Act of 1932* 'empowered a single commissioner to exercise powers in respect of main roads, developmental roads, tourist roads and secondary roads as a body corporate, the Department of Main Roads' or DMR (ibid.). At the same time as the DMR was created to handle road construction, the 1932 Act created a Department of Railways (to run the railway system) and a Department of Road Transport and Tramways (covering vehicle registration and licensing, and bus and tram services). Each had its own Commissioner. (New South Wales *Government Gazette* No.193, 29 Dec. 1932: 4591). Road traffic regulation (vehicle registration, driver licensing, collection of fees and taxes, traffic management, parking controls etc.) under the Metropolitan Traffic Act of 1900 was the responsibility of the police until 1930 when that function was handed to a Commissioner of Road Transport (and in 1932 to the Highways and Roads Transportation branch of the Transport Ministry and subsequently the Traffic Authority of NSW). Figure 5.2 below shows the development over time of the State bureaucratic agencies in the roads sector. As in Melbourne there was a process of consolidation of authority. The picture is complicated by the constantly changing political authority, the Ministry, but the evolution of the Ministry is treated separately below. Note that there is some overlap between the roads and public transport sectors since buses and trams run on roads: the Department of Railways and Department of Road Transport and Traffic (public transport agencies) are shown as dotted rectangles in Figure 5.2.

Up to 1931 the administration followed the 'professional bureaucratic model' (Laffin, 1995: 73) in which professional functional agencies had much autonomy in the conduct of their responsibilities. In 1931 and 1932 NSW Premier Jack Lang legislated to impose political will over the transport agencies and co-ordinate public transport and roads. A '*State Transport (Co-ordination) Act* was passed in 1931, only to be quickly replaced by the *Transport (Division of Functions) Act* of 1932, whose purpose was 'for the improvement and co-ordination of transport', and the *Ministry of Transport Act, 1932*. The latter Act created a transport 'portfolio' in which the Minister 'oversaw' the functional agencies (information obtained from the NSW State Records, http: //investigator.records.nsw.gov.au, accessed 7 April 2011).

This attempt at co-ordination of transport policy was conceived as a political strategy to protect the publicly owned railways from competition with privately owned buses and taxis. Broomham (2001: 119) reports that Lang: 'totally altered the administration of public transport and the positioning of roads and roads funding, and assumed control of main roads through the [NSW] *State Transport (Co-ordination) Act of 1931*. However, these arrangements were short lived as Lang was removed from office by the Governor General on 13 May 1932.

A further attempt at co-ordination was made in 1950 when the functions of the three separate transport departments (railways, road transport and trams, and main roads) were brought under the supervision of the NSW Transport and Highways Commission established by the NSW *Transport and Highways Act 1950*. According to the Act, this Commission was to supervise and coordinate the functions of the Ministry to provide a 'properly integrated system of transport'. The Commission's powers included reviewing policy for the development of transport throughout the State; making recommendations regarding proposals by the agencies, coordinating their budgets and 'controlling and directing' the agencies. (NSW *Transport and Highways Act 1950* s8 [1] [a]). This brought coordination of policy and finance for all transport modalities under one agency. This attempt also lasted a mere two years.

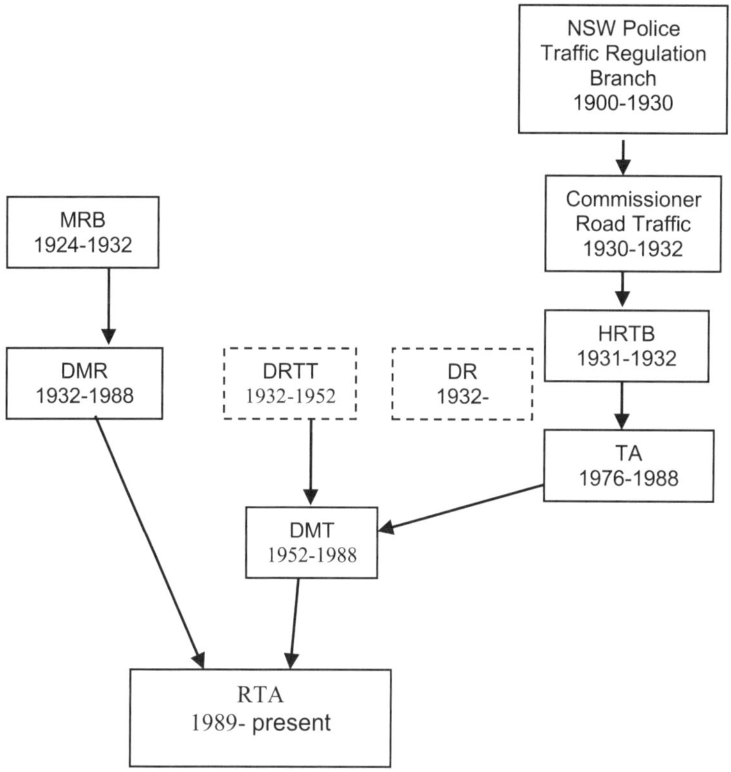

Key: DMR = Department of Main Roads; DMT = Department of Motor Transport; DRTT = Department of Road Transport and Trams; HRTB = Highways and Roads Transportation Branch; MRB = Main Roads Board; RTA = Roads and Traffic Authority; TA = Traffic Authority of NSW.

Figure 5.2 Sydney: Structural change in the roads sector

The co-ordination of transport in this period, though well intentioned, appears to have been little more than politically cosmetic, and real power continued to reside with the separate agencies. The 1950 Act is not mentioned by either the DMR in its own official history of itself (Department of Main Roads, 1976) or by Broomham (2001) from the perspective of the Roads and Traffic Authority. This serves to remind us that words in transport coordination (or integration) Acts do not necessarily mean that transport policy will in fact be coordinated or integrated in practice, or in the minds of key players.

In 1952, the Transport and Highways Commission was abolished, and the Department of Road Transport and Tramways was split into what became the Department of Motor Transport (vehicles) and the Department of Government

Transport (trams and buses) by two amendments of the NSW *Transport (Division of Functions) Act 1952*. (NSW *Government Gazette* No.114, 30 May, 1952: 1861). Effectively, each mode of transport had its own agency formally accountable to the ministers responsible for highways and public transport. From 1956 there was a separate Minister for Highways and Local Government.

In 1975, under a reforming Labor Government, a machinery of government review brought about the amalgamation of the ministerial portfolios for transport and highways, and the Ministry of Transport became the Ministry of Transport and Highways. Co-ordination was again attempted and the Ministry was given the task of, 'developing coordinated planning proposals and advising the Minister for Transport' (NSW *Public Service Board Annual Report*, 1975: 82). In following years the Ministry was charged with assisting the Minister in the initiation, implementation, evaluation and coordination of transport policy and development. Divisions included the Public Transport Commission, the Department of Motor Transport and the Traffic Authority of NSW. But the roads administration remained a separate domain from that of public transport. Figure 5.2 outlines the changing transport ministry from 1932 to the present time (2010).

The functions of regulating road traffic and building roads were also still separated. But from 1989 these functions were brought together under the NSW *Transport Administration Act 1988*. The Act amalgamated the Departments of Main Roads and Motor Transport to form the Roads and Traffic Authority. Responsibility for road construction, traffic regulation, driver licensing, and road safety became centralized within one agency, an arrangement which continues to the present day.

As in Victoria the span of control over the full range of functions relating to road planning, construction and management gradually became consolidated within a single agency, the RTA. While, as in Victoria, there have been repeated attempts throughout the second half of the twentieth century at co-ordination of all transport functions (not just roads), this has not diminished the integrity, independence or institutional power of the roads agency.

Perth

The principal State road planning, management and construction authority for Western Australia has remained largely unchanged since 1930. The Main Roads Department (MRD), later Main Roads WA (MRWA), was established by the WA *Main Roads Act 1930*: 'an Act to consolidate and amend the law relating to and making provision for the construction, maintenance and supervision of highways, main and secondary roads and other roads, the control of access to roads and for other relevant purposes' ('Preface' to the Act). The initial functions of the MRD were the construction and maintenance of main roads in the State. The MRD was to be headed by a single Commissioner of Main Roads appointed by the Governor in Council (which means essentially the State Cabinet) for a five year period. Under section 29(b) of the *Main Roads Act*, 'the Commissioner may enter upon, survey and compulsorily acquire land for the purposes of providing main roads'.

Amendments to the Main Roads Act introduced in 1996 widened the remit of the MRD to take on the broader role of 'roads manager and service provider' (DMR *Annual Report*, 1996). The amendments separated policy, standards and regulatory functions from service delivery, and also made reference to greater use of tendering and contracting out of services. But the overall integrity of the agency was not affected. The 1996 amendments provided for the establishment of a Main Roads Advisory Board (section 12) whose members were to be appointed by the Minister. This Board, however, has not been established as successive Ministers were not keen to take up the option (pers. comm. Martin, G., 21/01/2011). The Board's functions would have been to report in writing to the Minister about any matter referred to the Board by the Minister. The Board would also advise the Minister about the present and future needs of the State for roads and for infrastructure relating to road transport, and to advise the Minister of any concerns about the State's roads and infrastructure relating to road transport held by persons concerned with road transport or users of roads. Edmonds indicates that 'This was the type of Board that Don Aitken had tried to set up a decade earlier with a similar purpose: to bring the community formally into the process of providing road services'.[1]

The 1996 Amendments defined a remarkable power of the Department. Under paragraph 19B(3), the Minister *may not* give a direction with respect to the Commissioner's decision to make a recommendation to the Governor in Council under sections 13, 14, 24 or 28a, covering respectively the power to proclaim main roads, provide highways and main roads, proclaim secondary roads and controlled access areas. This section effectively gives the Commissioner the ability to override a decision by the Minister. Under the Act, the Governor may proclaim roads on the recommendation of the Commissioner (*Main Roads Act*, section 13[1]). The Commissioner is to report to the Minister when directed (section 19) and such directions are to be included in an annual report.

In the 1997 MRD Annual Report the purpose of the road expansion program is stated: 'to expand the road network to meet the present and future road transport needs of the community and to assist in the promotion of the economic growth of the State'. Thus road building and economic development are explicitly connected.

The Department also performs its functions through other Acts. The WA *Land Administration Act 1997* confirms the power of Main Roads WA to acquire land (Part 9 of the Act). The WA *Road Traffic Act 1974* provides a right of recovery for road damage (section 84), and credits funds from driving licenses to MRWA (section 22). The WA *Road Traffic Code* (Regulation 297) gives the agency the power over road signs. The WA *Road Traffic Vehicle Standards 2002* (Regulation 23) gives the agency authority to issue permits for heavy vehicles.

The MRD *Annual Report* (1987) describes a trend towards strategic planning 'to optimize the effectiveness and safety of the road system now and in future years'. This new focus is reflected in the appearance of a Strategic Planning section

1 http://www.ozroads.com.au/WA/history.htm accessed 21 January 2011.

in the agency's organizational chart from 1988 onwards. Evidence of a change in focus is illustrated by a comparison between statements of objectives in the annual reports of 1986 and 1989:

- 1986: To respond to Government policies, assist and cooperate with local government and other agencies.
- 1989: To provide Government with sound advice and support government strategies, cooperate with local government and public transport in developing and managing an integrated road network.

A major change occurred to the position of Main Roads in 2001. A Machinery of Government Taskforce under the leadership of Mr Stuart Hicks resulted in the creation of a consolidated Department of Planning and Infrastructure bringing together in one portfolio not only roads and public transport but also the town planning functions. This arrangement was designed to facilitate the aim of 'land use transport integration' and 'the integrated treatment of public transport modes and roads planning' (more will be said about the Hicks's initiative in the next chapter on public transport). The MRD was to become a service agency, with transport policy and strategy for roads falling to the responsibility of DPI. Of significance for the achievement of integration was the reduction from many Ministers with transport or land use responsibilities to just one Minister. Whether or not the DPI achieved integration in practice is debatable; a public forum held in 2008 gave some insights into the problems public officers experienced in attempting integration across the silos (Legacy et al., 2011). The DPI lasted until the Labor Government lost power in 2007. With the creation of the DPI, the Main Roads department lost some of its autonomy. However it remains a powerful agency. Like VicRoads in Melbourne and the Roads and Traffic Authority in Sydney, MRWA is a well-integrated and long lived organization with a strong focus, ample funding and considerable control over the span of functions to do with road planning, management and construction.

Conclusion

It is evident from our examination of how organizations shape road infrastructure that there is a debate around a set of key institutional factors each city is grappling with: integration versus fragmentation, mutual support versus conflicting perspectives, the autonomy of the agency and its relationship to the minister(s).

In Sydney, over time, structural change has seen the consolidation of roads agencies into one fairly autonomous roads authority as compared to the public transport structure (next chapter). Transport planning remains fragmented with roads and public transport agencies operating in separate compartments. Corporatization of agencies and the introduction of public-private partnerships increasingly prevent integration of transport planning.

In Melbourne the story is one of partial integration, thus setting it along similar lines to Sydney. The gradual consolidation of all roads-related functions within one stable organization together with the ability to leverage support from other players reveals that the roads organization has developed a strong and powerful capacity to influence the policy agenda, but roads planning has remained essentially segregated from the planning of public transport, and at arm's length from land use planning.

In Perth, by contrast, there has been a strong movement to integrate roads and public transport planning. In this chapter we see this in relation to the roads agency. Together with the next chapter, it will become clear that there has been a reduction in the number single purpose institutions (each with its own Minister) as the state moved to integrate the planning of all transport agencies. The examination of legislation, however, suggests that the road planning agency retains considerable power, despite organizational change seeking to reduce this. Perth has also gone further than either Sydney or Melbourne in its attempts to integrate transport planning with land use planning (discussed in detail in Chapter 7).

Chapter 6
How Organizations Shape Infrastructure: Public Transport Organizations

In this chapter we continue the examination of the organizational hardware of transport institutions in the three cities. We now turn to the public transport organizations. As in the previous chapter we look at the capacity of the organization to operate over the full span of functions relevant to its policy domain, and the organization's ability to influence related policy domains and access the highest levels of political decision making. As before, under each heading we set out the situation for each of the three cities in turn. In the final section of the chapter we conclude with a general comparison of the situation of the roads and the public transport agencies and the potential findings this comparison suggests for path dependence.

Structural Change and the Span of Operational Control

As with road infrastructure planning and management, so with public transport planning and management, the essentially mechanical systems of public transport require a 'mechanistic' structure with strong direction and planning by the centre. Citing Rothengatter, (2000) Nielsen et al. (2005: 50) write, 'In order to succeed in the urban transport market, the public transport system must offer a network of lines that ideally integrates all modes and lines for both intra-urban and interregional travel'. They argue that integrated planning of routes and timetables throughout the network is essential to competitive multi-modal public transport performance typified by the central control over service levels and timetables exercised by the Zürich Verkehrsverbund (see also Mees, 2005: 41). Integrated planning of key aspects of the task is therefore essential to effective performance in both roads and public transport planning.

Melbourne

In Melbourne's public transport governance during the last 100 years, a time of rapid change followed a long period of stability. This complex history is summarized in Figure 6.1 below. The first public transport agencies at the beginning of the twentieth century were organized on a single function basis. Following the construction of railways by private companies in the Melbourne metropolitan area from the 1850s and during the land boom of the 1870s and

1880s, the governance of metropolitan rail operations was given to the Victorian Railways Commissioners (VRC) established in 1883. The Commissioners oversaw the operations of the railways right up until 1973 when the VRC was restructured and expanded to include a seven member board in place of the three commissioners (Victoria *Railways [Amendment] Act 1972*). The successor institution, the Victorian Railways Board (VRB) continued the work of the VRC under the new arrangements.

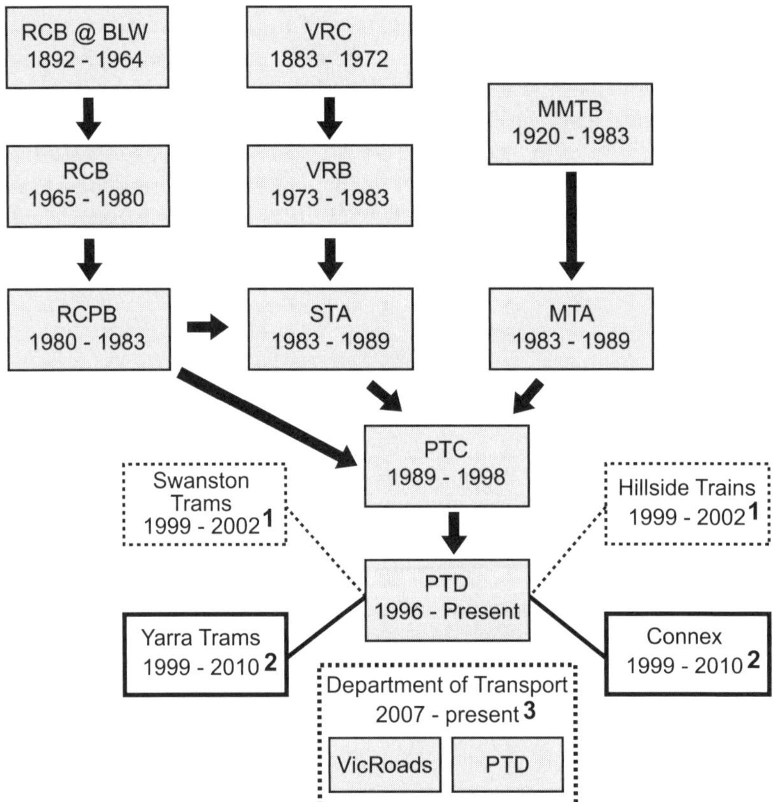

Key: BLW = Board of Land and Works; MMTB = Melbourne Metropolitan Tramways Board; MTA = Metropolitan Transit Authority; PTC = Public Transport Corporation; PTD = Public Transport Division (of the Department of Transport); RCB = Railways Construction Board; RCPB = Railway Construction and Property Board; STA = State Transport Authority; VicRoads = Roads Corporation of Victoria; VRB = Victorian Railways Board; VRC = Victorian Railways Commission.

Figure 6.1 Melbourne: Structural change in the public transport sector

The Railway Construction Branch at the Board of Land and Works was responsible for constructing railway lines. For much of the twentieth century, therefore, the building of railway lines and the running of train services were segregated. Construction was carried out by three single-purpose institutions before being merged into a multi-functional structure. The Railway Construction Branch was responsible for the construction of rail lines from 1892 until 1964 when it was abolished. All relevant responsibilities were then transferred to the newly established Railway Construction Board (RCB) which subsequently became the Railway Construction and Property Board (RCPB) in 1980. The last time a major piece of rail infrastructure was built – the Melbourne Underground Rail Loop – a special purpose authority was created to undertake it: the Melbourne Underground Rail Loop Authority (Melbourne Underground Rail Loop Act, 1970).

The Melbourne and Metropolitan Tramways Board (MMTB) was established under legislation enacted in 1918. The MMTB was responsible both for maintaining the tramways infrastructure and for providing a tram service (*Melbourne & Metropolitan Tramways Act 1918*). Bus services were run by private companies with the exception of the extensive bus service run by the MMTB (Tramway Buses). Originally Melbourne City Council (MCC) was the authority that licensed buses and routes in the Melbourne metropolitan area (a district that then extended eight miles from the boundaries of the City (Victoria *Motor Omnibus Act 1925*). Beyond the metropolitan area the Country Roads Board (CRB) was originally charged with the licensing of commercial passenger vehicles (Victoria *Motor Omnibus (Urban and Country) Act 1927*). When the Transport Regulation Board (TRB) was established in 1934 it inherited this responsibility from the CRB (Victoria *Transport Regulation Act 1933*).

The first comprehensive restructure of public transport commenced from 1982 with the election of the reforming Labor (social democratic) Government of John Cain junior. Following legislation in 1983 the MMTB became the Metropolitan Transit Authority (MTA) and gained operational responsibility for suburban trains. Railway construction and operational functions (excluding suburban trains) were united under the State Transport Authority (STA). This arrangement lasted only six years until the STA was merged with the MTA in 1989 to form the Public Transport Corporation (PTC). Together these moves, had they lasted, might have consolidated public transport operations and railway construction in the way that the roads planning and management tasks had been united within the Roads Corporation of Victoria (VicRoads) to make two equally strong transport agencies capable of planning and delivering new infrastructure. But the difference between the two was, of course, that the roads agency was not responsible for the operation of vehicles – this being the responsibility of individuals and firms – whereas the PTC was responsible for operating 'public' transport vehicles.

As patronage declined, public transport in the latter half of the century came to be increasingly subsidized from the State budget. Whereas few people questioned the private costs of running cars and trucks,[1] the public costs of running railways and trams was subjected to constant worried scrutiny, and reducing the 'burden on the budget' from public transport came to be an ongoing preoccupation of State governments.

In 1993 a reforming neo-liberal government decided to solve the budget 'subsidy' problem by privatization. In preparation for privatization, the operational sections of the PTC were divided into multiple business units, thus returning to a single function operational framework: Met Trains, Met Trams, Met Bus, V/Line Passenger (country services) and V/Line Freight. Then in 1999 the operations of the metropolitan tram and train systems were franchised out to four private consortia. The Public Transport Division (PTD) within the Department of Infrastructure (DoI) was created to administer the contracts, manage performance and undertake planning – but without the ongoing capacity to plan or enforce routes and timetables. Following the withdrawal of one of the international contractors (the UK company National Express, operating Hillside Trains and Swanston Trams) from tram and train operations at the end of 2002, each mode returned to the hands of a single operator (Connex, owned by Veolia, and Yarra Trams, owned by Transdev, both French transnational corporations).

Thus, for the first sixty-five years of its operation the tramways business was the responsibility of a single purpose institution. This stability was followed by 30 years of upheaval, including mergers with other public transport modes, splintering of amalgamated institutions and a shift from public ownership to private operation. Likewise, for 70 years the business of operating and constructing railways was managed by two separate single-purpose organizations. This period of stability was followed by 40 years of continual change, including integration of building and operating responsibilities, merging and dividing geographical accountabilities and shifting to a franchised contractual model for operations, which remains in place today.

Within the privatized system there is a role for the PTD to integrate public transport services. With increased pressure of demand for public transport, and the enactment of the *Transport Integration Act* (2009) this role is growing. Whether the PTD will take the step of integrating bus, tram and train services at the operational level remains to be seen. One potentially important initiative was the establishment of MetLink as a jointly owned initiative of private operators to ensure that information on services, timetables, routes and fares is provided to customers from a single source (MetLink, 2006). The franchise contracts provided for putative cooperation between all the public transport operators in the development

---

1 One who did forcefully raise the issue of the difference between road and rail, though from a different perspective, was Harold Clapp, the Chief Commissioner of Victorian Railways who warned that motor transport was a form of subsidized piracy 'since the motor bus, lorry or car travelled free on the public roads while the railways were obliged to pay for their own track (Davison, 2004: 126). Actually Clapp's statement points to an important difference in how trips are paid for which makes road based transport seem more attractive to travellers.

of timetables – '[t]he Franchisee must use its reasonable endeavours to achieve, on a consistent basis, the timetabled coordination of its Passenger Services with other services provided by those Operators in accordance with any guidelines issued by the Director' (section 7.11 of the Franchise Agreement – Trains and section7.10 of the Franchise Agreement – Trams). Thus two key advantages gained under the consolidated MTA structure are preserved under the privatized model, mediated through the PTD. There is a sense that public transport services need to operate as a single system, in contrast to the earlier competitive model in which each service responded only to its particular customers.

In response to national competition policy influenced by public choice theory, the 'mandarins' in charge of the public transport silos, after a brief period of organizational consolidation, were transformed into 'managers' and subjected to political direction (Self, 1993: 159; Capling et al., 1998: 84). In Melbourne this form of fragmentation took the form of privatization of service delivery with operational franchises under contract to the government. The result was the loss of the central control necessary to run a public transport system as an integrated network.

Sydney

The development of public transport organizations in Sydney is even more complex than in Melbourne. The trajectory of development is summarized in Figure 6.2 below. From 1932 responsibility for Sydney's public transport sector was divided between two State Government agencies: the Department of Road Transport and Tramways and the Department of Railways. In 1950 the NSW Transport and Highways Commission was formed in an attempt to control the various private and government public transport operators and integrate their planning with the main roads agency, but this attempt foundered after opposition from both private and government interests, and the 'entrenched institutions' (Rattray and Sinclair, 1978: 20–21). In 1952, the Department of Road Transport and Tramways was split into two separate departments: the Department of Motor Transport (vehicles) and the Department of Government Transport (trams and buses) by two amendments of the NSW *Transport (Division of Functions) Act 1952*. (NSW *Government Gazette* No.114, 30 May, 1952: 1861).

Twenty years later (1972), the two main public transport agencies were again brought together – this time separated from the roads agency – within a Public Transport Commission, which lasted until 1980. Rattray and Sinclair observed that the formal structures of the Commission enhanced the potential for the coordination of public transport decision-making. However its effectiveness, they note is vitiated by its paying less attention to the planning of public transport than to reducing the public transport deficit (the subsidy necessary to run a rail and bus system). The PTC, they state 'opposed or ignored attempts at metropolitan planning, generally making no overtures to planning bodies, and demonstrating an unwillingness to negotiate with other State instrumentalities or secure representation on bodies such as the NSW Traffic Authority' (Rattray and Sinclair, 1978: 22).

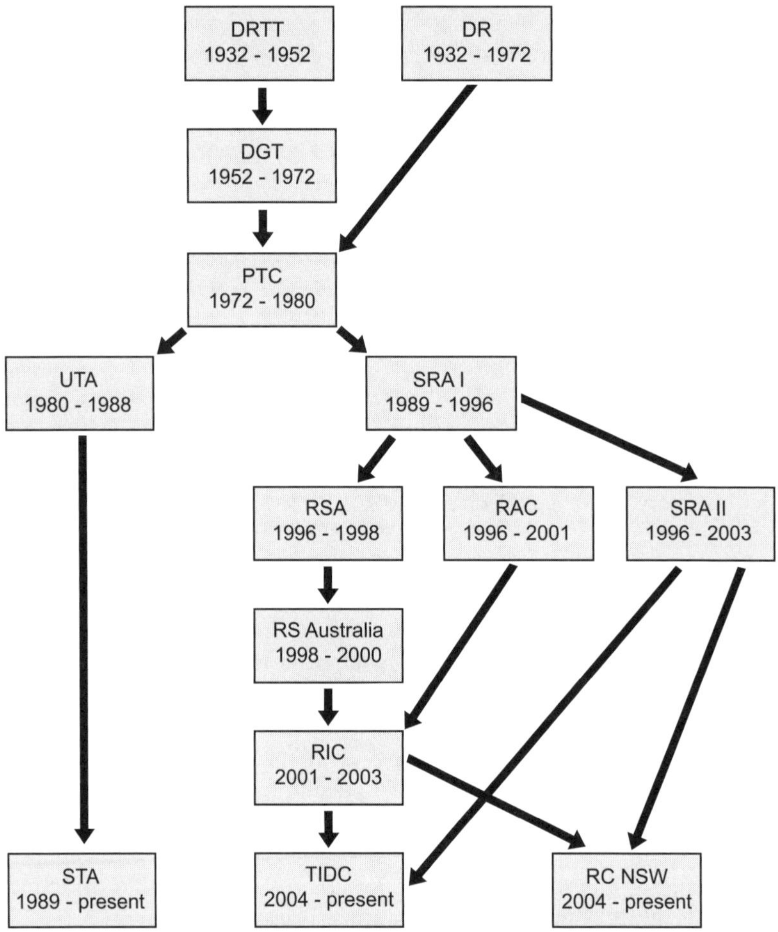

Key: DGT = Department of Government Transport; DR = Department of Railways; DRTT = Department of Road Transport and Tramways; PTC = Public Transport Commission; RAC = Rail Access Corporation; RC = Rail Corporation NSW; RIC = Rail Infrastructure Corporation; RS Australia = Rail Services Australia; RSA = Rail Services Authority; SRA = State Rail Authority; STA = State Transit Authority; TIDC = Transport Infrastructure Development Corporation; UTA = Urban Transit Authority.

Figure 6.2 Sydney: Structural change in the public transport sector

The late 1980s saw a further 'rationalization' of public transport, separating bus transport from the railways administration. The NSW *Transport Administration Act 1988* created the State Rail Authority (rail) and the Urban Transit Authority (bus) which soon afterwards became the State Transit Authority lasting until the present time. The Ministry of Transport became the Department of Transport in

1990 and was divided into three specialist areas (Strategic Policy dealing with the long term, Policy and Review dealing with current policy, and Service Planning and Coordination) In 1994, a further restructuring saw five divisions created to put increased emphasis on transport planning and rail safety. In 1995, the Department established a Transport Safety Bureau to improve the personal safety of passengers, the technical safety of the operations and to regulate the safety of industrial and commercial railways.[2]

In the 1990s, in response to the federal neo-liberal national competition policy, a period of corporatization took place in which a number of different railway businesses were identified and separated. Although the railways were retained in government hands, the corporations were expected to behave like successful private businesses – in readiness for privatization – whilst still having regard to social goals and compliance with the principles of sustainable development (*Transport Administration Amendment [Rail Agencies] Act 2003*, section 18B [2]). What followed was an extremely complex, fragmented and constantly shifting pattern of organizational development over nearly 15 years up to the time of writing (2011). The functions and life of the various railway organizations since 1989 is as follows. The State Rail Authority I (1989–1996) was created in 1989 under the *Transport Administration Act 1988***.** The State Rail Authority consisted of seven members and carried out the functions of the previous Public Transport Commission. The Authority operated through a Board under the direction of the Minister to 'efficiently operate all passenger and freight rail services, in conjunction with the Urban Transit Authority to provide efficient, economic and adequate urban passenger services' (State Rail Authority NSW *Annual Report 1984* p. 3). In 1996 The *Transport Administration Amendment (Rail Corporatisation and Restructuring) Act 1996* divided State Rail into four independent business units.

1. Rail Access Corporation (RAC) – owned the rail infrastructure and negotiated use of the track and funded upkeep.
2. Rail Services Australia (RSA) – function of maintaining railway tracks under contract to Rail Access, construction and repairs.
3. Freight Rail Corporation – commercial business providing freight services.
4. State Rail Authority II – provided city and country passenger rail services. It was also responsible for strategy and planning of services. (State Rail Authority *Annual Report, 1996*: 47).

This separation of 'businesses' was the beginning of the corporatization of rail in NSW which was supposed to foster competition in the marketplace.[3] Rail infrastructure was separated from rail operations in accordance with national

2 This move followed a series of rail crashes, the most recent up to that time being the Cowan disaster of 1990 in which six people were killed and 99 injured.

3 In 2004, the STA's Sydney ferries business was spun off as a state-owned company, Sydney Ferries Corporation.

competition policy, with freight services being corporatized into the 'Freight Rail Corporation' (Freight Rail, later FreightCorp) in preparation for eventual privatization. The 'Railway Services Authority' (RSA 1996–1998) was established with responsibility for track maintenance and construction activities (much of which was contracted out to private companies). The Rail Access Corporation (RAC, 1996–2001) was established under the Act with responsibility for managing the rail network and administering access by public and private operators. Only four years later the RSA and RAC were amalgamated to form the Rail Infrastructure Corporation under the *Transport Administration Amendment (Rail Management) Act 2000*. One of the primary objectives of the RIC was 'to operate as a successful business' (Schedule 2, section 2.1, [12] of the Act).

Just three years after that (in 2003) another change occurred. The functions of the metropolitan sections of the RIC and the passenger train operations were merged to form the Rail Corporation of New South Wales. The Corporation was established under the *Transport Administration Amendment (Rail Agencies) Act 2003*. RailCorp has the power to establish new rail services, own, manage, and establish rail infrastructure facilities, and acquire land to do so (NSW *Transport Administration Amendment (Rail Agencies) Act 2003* s6–11). As a state owned corporation RailCorp receives direction from the transport minister. The Board consists of between 3 and 7 Directors and is appointed by the voting shareholders. The same Act of 2003 established the Transport Infrastructure Development Corporation (2003–present). The objectives of the corporation were: 'to develop major railway systems' and 'to develop other major transport projects in an efficient, effective and financially responsible manner' (section 18B [1] of the Act). Three years later responsibility for integrated ticketing across modes of public transport was handed to yet another (unintegrated!) corporation, the Public Transport Ticketing Corporation (2006–present), something the more integrated PTC established in Melbourne in 1998 (Act no. 5, 2006, NSW, *An Act to Amend the Transport Administration Act 1988* to constitute a Public Transport Ticketing Corporation. See www.parliament.nsw.gov.au).

Manning (1991: 95) writes that the administrative changes of the 1980s 'resulted in the creation of an exceedingly large and unwieldy organization which would have been difficult to weld together at the best of times'. Subsequent history shows that the situation did not improve. While the roads institutions were consolidated and logically structured from 1989 with the creation of the Roads and Traffic Authority, the public transport institutions underwent a long period of change, fragmentation, and instability which continues to the time of writing.

Perth

Perth can now boast a better organized, stronger and more integrated administration of public transport than either Melbourne or Sydney, and this is demonstrated in the relatively efficient delivery of public transport networked services (see Stone, 2009). But the organizational history of public transport administration

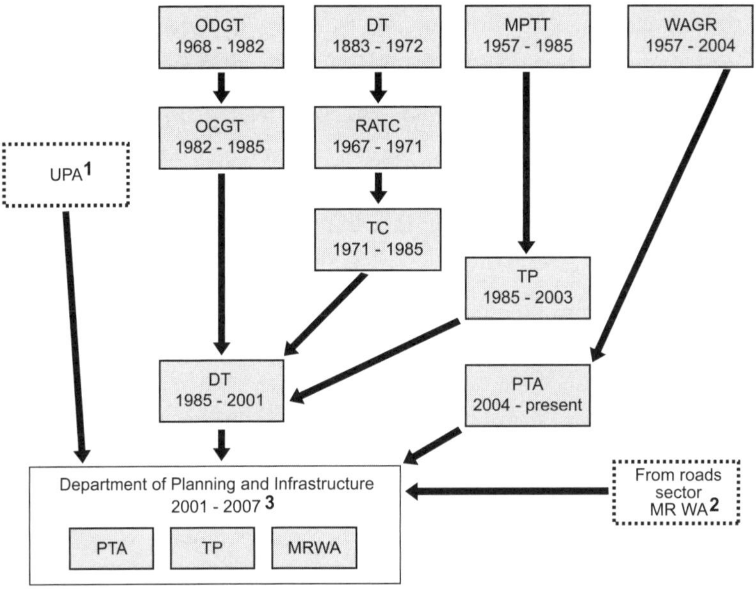

Key: DT = Department of Transport; DPI = Department of Planning and Infrastructure; MPTT = Metropolitan Passenger Transport Trust; MRWA = Main Roads Western Australia; OCGT = Office of Coordinator General of Transport; ODGT= Office of the Director General of Transport; PTA = Public Transport Authority; RATC = Road and Rail Transport Commission; TC = Transport Commission; TP = TransPerth; UPA = Urban Planning Administration; WAGR = Western Australian Government Railways.

Figure 6.3 Perth: Structural change in the public transport sector

began by following a familiar pattern; trains, trams and buses came under separate administrations (see Figure 6.3).

The Western Australian Government Railways Commission (WAGR) was established under the *Government Railways Act 1904*. The railways were the responsibility of the Minister for Railways until 1974 and The WAGR was a body corporate whose functions included making policy decisions with the approval of the Minister for Railways. In 1957, the *Government Railways Act 1957* sought to update and amend the law relating to the maintenance and management of government railways 'and for connected purposes'. The Commissioner was to manage railways subject to this Act and appoint officers without the need for Ministerial approval. Provision was made for quarterly reporting and for regulations to be gazetted.

In the 1950s, buses were provided by the private sector and a number of private operators. The Metropolitan Passenger Transport Trust (MPTT) was created in 1957 to allow for the merger of a number of private bus companies and government tramways. Buses and trams then became the responsibility of the MPTT, a Board established under the WA *Metropolitan (Perth) Passenger Transport Trust Act 1957*. At that time, therefore, the MPTT and the WAGR functioned quite autonomously with general direction from their respective Ministers and with responsibility for implementing policy decisions. Thus: 'When the MPTT was formed, emphasis was placed on the new body's commercial autonomy, reflecting the amalgamation of the various private bus companies which were taken over by the Government when their financial position worsened. The establishing Act of Parliament gave the MPTT the power to act largely independently of the Minister responsible for it (MPTT *Annual Report, 1985*).

In 1974, the responsibility for the policy and financial operations of the WAGR was taken over by the Minister of Transport (WAGR *Annual Report, 1974*). Under the *MPTT Amendment Act* 1974, policy control of all suburban public passenger transport was vested in the MPTT. Operational control of rail services remained with the WAGR but policy direction and financial responsibility was assumed by the Trust (WAGR, *Annual Report 1974,* p. 13). The reason for this transfer of responsibility was given as: 'The necessity to operate as an instrument of government policy and as a "public service" continues to inhibit the Railways in its endeavours to operate commercially in an increasingly competitive transport environment' (WAGR *Annual Report, 1974*). This move placed the responsibility for all suburban passenger services under the one authority, the MPTT and a single Minister for Transport. At that time the Government *Railways Act* was amended to reflect the change. With the Commissioner of Railways now a member of the Trust, 'this was considered to be a very workable arrangement which opens the way for full integration of the two services' (WAGR *Annual Report, 1974*). At that time, as patronage of the passenger services was falling, the WAGR was responsible for the provision of freight services (a major responsibility, with the State covering such a vast area and serving key agricultural and mining communities) which provided much of the needed funds to run a railway system. It would appear that the need to be accountable for the federal funds being awarded to public transport and the falling rail passenger patronage resulted in a reduction in power for the WAGR with the loss of the Minister for Railways and increased accountability for finances.

The mid-1980s marked a key change in the organization of public transport in the State. The MPTT in 1984 argued that public transport over the last 27 years had become increasingly viewed as a social service and not just a commercial operation providing transport to customers. Bus patronage as well as passenger train patronage continued to fall (MPTT *Annual Report, 1984*). The *Metropolitan (Perth) Passenger Transport Trust Amendment Act* 1984 came into operation, in addition to enlarging the Trust's Board from 4 to 7 members, the Act 'formally brought the MPTT under the general control of Government through the Minister for Transport'.

At this point it is worth noting that within the portfolio of the Minister for Transport were three continuous entities: The Department of Main Roads which became Main Roads WA and continues as a separate entity today, The Department of Transport (1961–1967), succeeded by the Road and Air Transport Commission

(1967–1971), succeeded by the Transport Commission (1971–1985), and the Metropolitan Passenger Transport Trust (1957–1985). All of the public transport entities (but not the Main Roads department) were amalgamated in one Department of Transport in 1985 which lasted until 2001. With this amalgamation the MPTT changed its corporate identity with a new name *TransPerth*, a new logo, and a new livery for its buses. The change was intended to give new branding and marketing strength to the bus services and did not entail legislative or administrative change. The head of the Department of Transport reported to the Minister for Transport who had parliamentary responsibility for the Department, TransPerth, the WAGR and Main Roads.

In 1993, taking into account national competition policy and the federal demand to foster competition in the provision of public transport services, the Minister of Transport stated:

> I do not believe that we should return to the pre-1958 arrangement of uncoordinated privately owned public transport operators. Our enthusiasm to become more efficient and competitive must be balanced with the continuing need for a well-coordinated and integrated transport service based on buses, trains and ferries and using the particular strengths of each mode in a complementary way. (MPTT *Annual Report, 1993*: 5)

This statement signalled a real difference in the provision of public transport from that pursued in both Victoria (Melbourne – privatization) and New South Wales (Sydney – corporatization). Integration of bus and train services from this point on became something more than aspirational. In 1994, it was reported, 'Considerable progress has been made with the transfer of planning and coordination functions to the DoT and preparing the MPTT to meet potential competitors consistent with the Government's desire to restructure public transport operations and open up the system to competitive tendering for service provision' (MPTT *Annual Report, 1994*: 40).

In 1999 the DoT announced, 'new directions in management emphasizing the contracting out of services which can be better managed by the private sector … it is transport's challenge to ensure continued deregulation … In 1999–2000 we will be seeking even more private sector involvement in the provision of transport services and place more emphasis on community and industry involvement in transport planning and decision-making'. (MPTT *Annual Report, 1998*: 3). Thus by the end of the 1990s there were already moves towards integration of bus and train services and to establish a model of contracted bus services with central control of routes and timetables.

As noted in Chapter 5, a significant development affecting the institutional structure of transport agencies occurred in 2001. The Labor Government elected in 2001 appointed a 'Machinery of Government Taskforce' to review the current arrangements for transport and make recommendations for the clarification of hierarchy and relationships between the transport entities within Government. The work of the Taskforce was led by man with a long history of involvement with the transport institutions in Western Australia, particularly the bus system, Mr Stuart Hicks. The 'Hicks Report' (Hicks et al., 2001) resulted in the WA *Machinery of Government (Planning and Infrastructure Amendment) Act 2001* (MGPIA Act 2001) which saw significant structural change. The Act created a new Planning and Infrastructure portfolio. As a result the Department of Transport (DoT) was abolished and the Department of Main Roads was moved into the new portfolio and in some ways demoted to become a 'service agency'. All of the functions of the DoT with the exception of road safety were transferred to the DPI on 1 July 2002. The MGPIA Act 2001 enabled the *Transport Coordination Act 1966* and the *Road Traffic Act 1974* to be brought under the Planning and Infrastructure portfolio.

The Hicks Report made recommendations for the hierarchical structure of the new portfolio and prescribed the various functions in the hierarchy. The resultant structure can be seen at Figure 6.3. The arrangements comprised one Minister with responsibility for the whole portfolio (a reduction from the recent past where there was a Minister for Planning and a Minister for Transport). The DPI essentially existed to support the Minister in pursuing the Government objectives and delivering services. The arrangement provided for direct reporting and assumed a close working relationship between the Head of Department and the Minister. The intent was for the Department to have minimal financial autonomy with dependence on the Consolidated Fund for its operation and viability. This enabled the Government closely to control the priorities of and resources allocated to the organization. Such Departments of State are normally characterized by a desire for staff to be employed under the WA *Public Sector Management Act 1994*. Within the portfolio the Main Roads Agency and the Public Transport Agency become 'service' or 'operating' agencies within the DPI, although it can be seen (in Figure 6.3) that the eventual arrangements were somewhat different. The intent of the Hicks Report was that road safety would also move to DPI but a late decision to suit the government's Ministerial portfolio saw the introduction of the WA *Road Safety Council Bill 2001* which gave the Road Safety Council its own legislation and so enabled the Office of Road Safety be transferred to the Department of Premier and Cabinet.

The PTA took on the operational functions of transport services, infrastructure development, ticketing and all other aspects of public transport. The *Annual Report* (2005: 1) describes PTA's purpose as to provide: 'accessible, reliable and safe public transport systems; protection for the long-term functionality of the rail corridor and railway infrastructure'. These systems consist of a bus network, an electrified suburban train system and a ferry service. Routes and timetables for the metropolitan system are planned and managed by the TransPerth Division and the Regional and School Bus Services Division of the PTA. TransPerth bus and ferry services are provided by private sector operators under commercial contracts, while TransPerth train services are provided by government via TransPerth Train Operations Division of the PTA.

The PTA is required to carry out planning of the public transport system. The PTA Act 2003 is headed by a Chief Executive Officer appointed under the WA *Public Sector Management Act, 1994*. Under this Act, the accountability of the PTA is more strictly prescribed than any of the other institutions. For example, under section 20, 'In each financial year the Authority is to prepare, and submit to the Minister for approval, a draft operational plan'. The contents of the operational plan are prescribed and include: objectives, business and performance targets, resource use, an outline of the nature and scope of functions, and initiatives to improve consumer outcomes. The plan can only be modified with Ministerial approval and concurrence is required from the Treasurer. The Authority is to strictly adhere to the plan. The PTA must consult the Minister before it enters upon a course of action 'that in its opinion, (a) amounts to a major initiative; or (b) is likely to be of significant public interest' (section 25 of the Act). In addition to its responsibilities under the PTA Act 2003, the PTA is given powers under the WA *Government Railways Act 1904*. Part III of the Act gives the authority (the PTA) the direction, management, maintenance and control of the Western Australian Government Railways subject to (c) the *Railways (Access) Act 1998*, the *Rail Freight System Act 2000*; and (d) the *Rail Safety Act 1998*.

Governmental arrangements change frequently over the electoral cycle, so also the lines of accountability to one or more ministers, and the names of departments. What is important about Perth public transport is the steadily increasing control by a single agency over the full span of the functions of public transport. The PTA therefore is now able to carry out a powerful planning function for public transport in a way that the roads agencies in all three cities plan for roads construction and management.

Discussion and Conclusion

A comparison between the roads and public transport organizational sectors is revealing of their respective capacities to deliver and implement coherent policies (Astle, 2007). The advantage of a single, multi-purpose institution is that it can ensure that all transport initiatives are consistent with the corporate vision (both in the roads and public transport sectors). Also, internal conflict can be avoided as any negative consequences generated by one policy proposition can be assessed for their impact on other policies, and corrected. Potentially conflicting priorities within the sector can be sorted out prior to adoption of policy.

Another advantage of consolidating all activities within a single institution is the ability to harness efficiencies, both in corporate and customer services. For example, in the roads sector in Victoria the capacity to develop regional offices for vehicle registration and licensing would not have been possible when the Motor Regulation Branch (Victoria) was a separate institution, as sufficient funds could not be found (Anderson, 1990: 119). An additional benefit of merging roads, and or, transport institutions is that it creates a strong influence on wider transport and urban policy. Teams can be assembled to generate policy solutions which approach the problem from multiple technical perspectives. This is certainly the case in Perth where the creation of the Department of Transport provided the opportunity for a stronger public transport voice. Combining knowledge and expertise across functions creates a well-coordinated response from the sector. For example VicRoads has made regular submissions to inquiries conducted by the Joint Parliamentary Committee on Road Safety (Victoria).

In all three cities the roads agencies remain integrated, powerful and well-resourced units within the departmental structure. In contrast the public transport agencies have moved from broad ministerial control to a stronger connection between the service delivery agencies and the bureaucratic level. In Sydney and Melbourne service delivery is separated from strategic policy and fragmented (corporatized or privatized). In Perth the state agency retains strong contractual control over the service delivery despite the fact that bus services are delivered by private companies.

Institutions in the roads sector had early access to a regular conference of contemporaries, and have subsequently gained access to a national forum as well as initiating a cross-sector group to consider road management issues (Astle, 2007). The public transport sector, by contrast, has had access to only one national association, the Australasian Railways Association, which has historically placed most emphasis on national railways in Australia and New Zealand (and in particular inter-state freight operations rather than passenger rail). There is no peak national body dealing with urban public transport across modes. The public transport sector has thus had weak opportunities to leverage the support of organizations with common interests.

We can see, therefore, that whilst for both sectors the number of additional actors with an interest in the industry has increased over the last 50 years, the roads sector has emerged with a greater level of support from a range of interest groups including business and trades union interests involved in road construction and management, traffic and transport consultants involved in modelling demand, the Australian Road Research Board and the road traffic service agencies like the NRMA in Sydney and the RACV in Melbourne. Compared with the public transport sector, players in the roads sector have significantly more resources at their disposal and display a greater degree of mutually compatible and supportive interests. For instance the RACV advocates on behalf of the motoring community for better infrastructure and safety initiatives, the ARRB Group and the BITRE contribute relevant research, civil construction firms want to win the resulting business contracts, consultants are constantly called upon to conduct well-funded demand and benefit cost studies, and is the roads agencies are responsible for planning, developing and maintaining the road network. Being able to harness a number of voices and resources in support of their initiatives places the roads agencies in a powerful position to influence and determine the policy agenda.

In summary, the roads agencies, in a period of strong growth in road traffic, have been able to gather and build support for their policy preferences, sharing ideas with sister organizations in other States to lobby for additional funds from the federal level. By contrast public transport agencies, facing declining patronage, have been more locally focused on the problem of the so called 'public transport deficit', and historically have had limited opportunities to engage with other players, which has restricted their ability to leverage support for policy ideas.

In the next chapter we turn to the influence of the organizations in the two sectors on transport plans, and the impact of those plans on the development of infrastructure.

Chapter 7

Transport Plans in Melbourne, Sydney and Perth

In the previous two chapters we showed how the organizational hardware of the roads and public transport sectors developed different potential capacities to influence policies and plans. In this chapter we examine the plans themselves and their outcomes. The chapter is structured to place the greatest emphasis on critical junctures in transport planning for the three cities, with a particular focus on the most recent period from the late 1990s to the present in which a new storyline of 'integrative planning' emerged in which sectoral plans (for instance for different modes, or separate land use and transport domains) were replaced by plans that attempted to plan for the transport system as a whole, including its connection with land use development and regulation.

The term 'integrative' is chosen rather than 'integrated' because, at least in Melbourne and Sydney, it is not yet possible to say that transport and land use plans have actually resulted in integration of transport modes, let alone integration of transport and land use plans. Rather there emerged an aspiration to integration that became more and more insistent, and it is this aspiration we term 'integrative'. First though, for each city we outline the trajectory of planning from the 1950s up to the turn of the century. We comment on these plans only briefly, highlighting key moments, since they will be further discussed in Chapters 8 and 9 in relation to the discourse of transport policy.

Melbourne: from Roads Dominance to Integrative Planning

The organizations in the roads sector have been continuously involved in the metropolitan planning process for much of the last 50 years. The CRB contributed to three key transport plans: The 1969, 1978 and 1980 plans (see Table 7.1 for details).[1] The making of these plans also included representatives from the public transport sector; however the influence of the roads agency appears to have been stronger in shaping the policy agenda.

---

1 The CRB was not involved in the 1954 Metropolitan Plan, primarily because at the time responsibility for metropolitan roads was vested with municipal councils before being transferred to the MMBW in 1956. The CRB did not gain responsibility for metropolitan arterials, freeways and bridges until 1974 (*Metropolitan Bridges Highways and Foreshores Act 1974*).

Table 7.1 Key transport and land use plans for the Melbourne metropolitan region

Year	Plan	Authority and representation
1954	*Metropolitan Planning Scheme*	Melbourne & Metropolitan Board of Works
1969	*Melbourne Transportation Study to 1985*	Country Roads Board, Traffic Commission, Transport Regulation Board, Melbourne & Metropolitan Board of Works, Victorian Railways Commissioners, Melbourne & Metropolitan Tramways Board, Minister of Transport, Coordinator of Transport and three non-transport representatives*
1971	*Metropolitan Planning Scheme*	Melbourne & Metropolitan Board of Works
1978	*Transport Plan*	Country Roads Board, Road Safety & Traffic Authority, Victorian Railways Board, Director of Transport and eight non-transport representatives**
1980	*Victorian Transport Study*	Country Roads Board, Transport Regulation Board, Victorian Railways Board, Melbourne & Metropolitan Tramways Board and W M Lonie
1981	*Metropolitan Strategy Implementation*	Melbourne Metropolitan Board of Works
1987	*Metropolitan Arterial Road Access Study*	Road Construction Authority, Road Traffic Authority and Ministry of Transport
1987	*Shaping Melbourne's Future*	Department of Planning and Environment
1988	*MetPlan*	Metropolitan Transit Authority and Ministry of Transport
1993	*Public Transport Reform*	Minister for Public Transport
1994	*Linking Melbourne*	VicRoads
1998	*A System into a Service*	Department of Treasury and Finance
2002	Melbourne 2030	Department of Infrastructure
2004	Linking Melbourne: Metropolitan Transport Plan	VicRoads and Department of Infrastructure
2006	Meeting Our Transport Challenges	VicRoads and Department of Infrastructure
2008	Investing in Transport (East West Link Needs Assessment)	Department of Infrastructure (East West Link Needs Assessment conducted by Sir Rod Eddington)
2008	*The Victorian Transport Plan*	Department of Transport
2009	*Melbourne @ 5 Million*	Department of Planning and Community Development
2009	*Towards an Integrated and Sustainable Transport Future: A new legislative framework for transport in Victoria*	Department of Transport

* Minister for Local Government, City of Melbourne and Director of Finance.
** *Prem*ier's Department, State Co-ordinator of Works, Director of Conservation, Town & Country Planning Board, Director-General of Education, Secretary for Lands, Secretary for Planning, and the Melbourne & Metropolitan Board of Works

All the key transport players were involved in the drafting of the plan that emerged as the 1969 Melbourne Transportation Plan (Metropolitan Transportation Committee, 1969). The constitution of the Committee reveals that the roads sector had the most actors and thus potentially the determining voice. The two public transport representatives were outnumbered by the five roads representatives, and the remaining members were split among a range of single interest actors. The Committee recommended the construction of a vast metropolitan freeway grid taking up 85% of the total projected expenditure. This plan was incorporated into the 1971 MMBW plan for Melbourne, which represented the zenith of the vision promulgated by the roads organization of a Melbourne modelled on the American ideal of free movement for private vehicles over the whole metropolis. Its partial abandonment was a major setback for the thrust of roads planning. As Lay (2003: 201) put it, 'The new Premier announced the end of all dreams of a Melbourne enmeshed in a web of freeways when he declared that the planned network had been cut in half on sociological and environmental grounds'. Nevertheless the dream continued to influence the pattern of road infrastructure that was relentlessly pursued wherever possible by the roads agencies. Its impact on Melbourne's developing infrastructure can be seen in Figure 7.1.

The 1978 plan recommended options for the whole of Victoria over the next five years and also made proposals to be considered for the longer term future. Recommended public transport improvements included: completion of the underground loop, continued replacement of rolling stock in all modes, separation of tram and motor vehicles where applicable and construction of transport interchanges at Box Hill and Frankston including grade separations. However this plan did not presage a turn towards priority for public transport. A study of transport was commissioned to be chaired by Mr W. M. Lonie (a former car company executive). The Lonie report strongly criticized the public transport sector for the high level of government subsidies received, declining patronage and perceived operational inefficiencies. The report recommended cutting routes and services with low levels of patronage, conversion of tram services to bus operations and the introduction of efficiency measures (Victorian Transportation Study, 1980a: 100–103). In contrast, the roads sector was praised for its increased and popular usage.

While in the first periods of multi-sector planning the roads agencies and their agenda were strongly dominant, from around the turn of the century public transport began a slow recovery of its influence in parallel with an increase in patronage that surprised those in power. The first sign of such planning came with the publication in 2002 of the government's land use and transport plan, Melbourne 2030 (see Government of Victoria, 2002, prepared within the Department of Infrastructure). Notable for its adoption of an 'urban growth boundary', 'a more compact city', and a large number of scattered suburban 'activity centres', the plan set a goal of increasing public transport's mode share of motorized trips in Melbourne from 9 per cent in 2002 to 20 per cent in 2020 (ibid.: 146). While the plan talked about upgrading and developing the 'principal public transport network' (Policy 8.1),

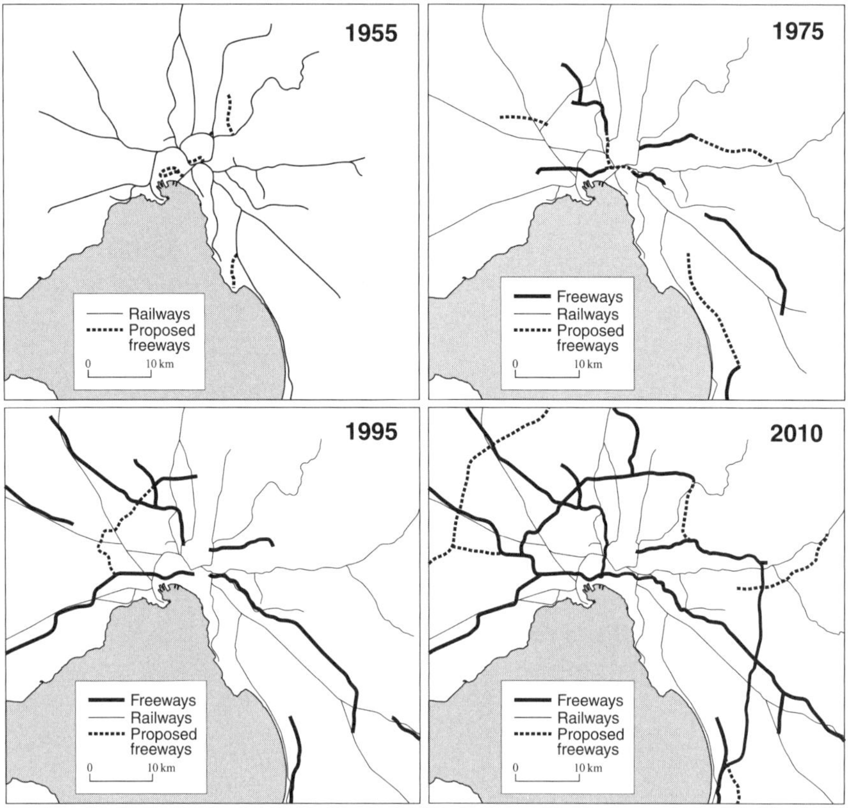

Figure 7.1 Melbourne: Railways and freeways 1955–2010

there was no specific program of action to achieve the ambitious 20–2020 goal. However there was a distinct change of emphasis in favour of public transport, and a number of modest projects were mooted (for example a small extension to the 109 tramline (ibid.: 151), and improvements to bus and tram services (ibid.: 149). There was a commitment to 'give more priority to cycling and walking in urban development' (Policy 8.7). The aspiration 'to coordinate all transport modes to provide a comprehensive transport system' was stated (ibid.: 145). The Plan incorporated one large scale road project, the Scoresby freeway (subsequently EastLink) through Melbourne's outer eastern suburbs.

There followed a series of integrative plans in which public transport was increasingly recognized as important to the functioning of the city. The *Linking Melbourne: Metropolitan Transport Plan*[2] appeared to be the combined work

2 This plan is a more detailed articulation of the vision set out in *Melbourne 2030*, which is a blueprint for the future of Melbourne. The key players in developing this strategic

of VicRoads and the Public Transport Division (PTD) at the Department of Infrastructure. The transport policy problems were articulated and then matched with both road and public transport solutions. In May 2006 a new report emerged, *Meeting Our Transport Challenges*. New metropolitan public transport projects included creating a cross-town bus network, boosting capacity on the rail network, adding more train and tram services, improving safety and promoting the use of public transport. The total estimated cost of these initiatives was $6,515 million, which represented 62% of the whole plan. The roads proposals included building better connections to outer suburbs and delivering a better east-west link. The estimated cost was $3,445 million, approximately 33% of the total.[3]

Meanwhile the government was in process of completing two major road projects: the 'Scoresby Transport Corridor' which became EastLink – a new tollway built as a private-public partnership (AU$2.5 billion) and an element in Melbourne's ring road, the enlargement of the Monash – CityLink – Westgate motorways (AU$1.39 billion, later blowing out to well over AU$2 billion), the Deer Park bypass in Melbourne's west (AU$331 million) and AU$83.5 million of other urban road improvement projects (see Government of Victoria 2008: 50, 92).

Three further transport planning reports were published in the later 2000s: the Report of the East West Links Needs Assessment conducted by consultant, Sir Rod Eddington (2008), the Victorian Transport Plan (2008), and the report on proposed new transport legislation (2009). These continued the theme of integrative transport planning, with a continuing thrust towards road building tempered by commitments to increased future expenditure on public transport. The latter were no doubt driven by both a dramatic increase in public transport patronage since 2004 and continual expressions of public dissatisfaction with the system.

The Victorian Transport Plan (2008) continued to support future large scale investment in the public transport network, though it appears that the roads agency inserted some major new extensions of Melbourne's motorway system (further demonstrating the continual 'behind the scenes' planning of road projects). The gradual conversion of Melbourne's creaking suburban railways into a modern metro system was mooted, and importantly 'behind the scenes' planning of new public transport infrastructure was suggested: 'Planning for the development of new rail lines in the future is underway to develop and preserve options for a range of future rail lines including the new growth areas of Clyde, Mernda, Aurora and North Epping'. In part this statement is a response to the repeated calls by rail advocates to bring rail connections to the outer suburbs, together with reluctance by the government to do so. Only minor extensions of the rail system to Cranbourne East in the south-eastern suburbs and South Morang in the north were included in the Plan. The balance of expenditure in the Plan between road and public transport

---

document included the Department of Premier & Cabinet, the Department of Sustainability & Environment and the Department of Infrastructure.

3 The remaining 5% is set aside for public transport projects in regional Victoria (GoV, 2006).

investment again favoured the latter: AU$16.23 billion of public transport related projects, and $12.31 billion of road related projects (including AU$440 million on grade separating road-rail crossings). A small amount (AU$100 million over a period of ten years or more) was allocated to improving cycling infrastructure.

A final institutional step in integrative planning was with the introduction of a new legislative framework for transport governance and management in the Transport Integration Act (2010). The proposed legislation sought to create a common framework to guide future transport planning, incorporating a 'vision statement', 'transport system objectives' and 'decision-making principles'. The Vision statement was aspirational:

> Victoria aspires to have an integrated and sustainable transport system that contributes to an inclusive, prosperous and environmentally responsible state. (ibid.: 12)

The system objectives stated as (in this order): efficiency, coordination and reliability, safety, health and wellbeing, economic prosperity, integration of transport and land use, social and economic inclusion, and environmental sustainability. The Act does, as claimed, provide 'strong legislative support for the development of an integrated and sustainable transport system' (Transport Integration Act, Victoria, 2010 accessed 7 December 2011), yet the Act does not specify how such a system is to be achieved. It advocates integration but does not itself undertake it. In the end the Act is, as it says, 'aspirational'. It remains to be seen whether and how integration actually occurs. Up to now what is most evident is a more equal balance of competition for investment projects between roads and public transport.

Sydney: Road, Rail and Integrative Planning with Political Instability

The first major metropolitan planning exercise for Sydney (The Royal Commission of 1908–1909) was much concerned with traffic congestion, but significantly the solutions at that time were mainly rail-based (Ashton, 1995: 38). The emphasis, however, shifted after World War 2 from railways and public transport to roads (see Table 7.2).

In the 1950s the Department of Main Roads, influenced by the development of highways in the United States commenced a program of reserving land for the future construction of arterial roads throughout the region (Myer, 1993: 211). The plan, known as the Main Roads Development Plan, was the first long-term plan for freeways and other major urban roads in Australian cities (Ball, 1996: 11). Metropolitan planning began with the enactment of the Local Government (Town and Country Planning) Amendment Act of 1945 creating the Cumberland County Council with jurisdiction over the whole of metropolitan Sydney (Sandercock, 1975; Stretton, 1975). The DMR's Deputy Chief Engineer became the first Chief

Table 7.2 Key transport and land use plans for the Sydney metropolitan region

Year	Plan	Authority and representation
1951	*Main Roads Development Plan*	Department of Main Roads
1948–1951	*Cumberland County Planning Scheme*	Cumberland County Council
1968	*Sydney Region Outline Plan*	State Planning Authority
1974	*Sydney Area Transportation Study*	Ministry of Transport
1987	*Roads 2000*	Department of Main Roads
1988	*Sydney into Its Third Century: Metropolitan Strategy for the Sydney Region*	Department of Planning
1993	*Sydney – Cities for the 21st Century*	Department of Planning
1993	*Integrated Transport Strategy for the Greater Metropolitan Region*	Department of Transport
1997	*A Framework for Growth and Change: The review of strategic planning the greater metropolitan region*	NSW Department of Urban Affairs and Planning
1999	*Action for Transport, 2010 – An Integrated Transport Plan for Sydney*	Department of Transport
2005	*Metropolitan Strategy: City of Cities – a Plan for Sydney's Future 2005*	Department of Planning
2005	*Metropolitan Rail Expansion Plan*	Department of Transport
2010	*Metropolitan Transport Plan, Connecting the City of Cities*	Department of Transport and Infrastructure
2010	*Metropolitan Strategy Review*	Department of Planning
2010	*Independent Public Inquiry: Long term Public Transport Plan for Sydney*	Independent Public Inquiry

County Planner. The County planners utilized the DMR's previous studies, including detailed land-use and population density data, and incorporated the main features of the Main Roads Development Plan into the Cumberland County Plan.

Under attack from various quarters, the County Council was finally abolished in 1963. Its successor was the State Planning Authority (SPA), 'a predominantly technical and professional body' with the addition of five local government representatives (Rattray and Sinclair, 1975: 27). The roads strategies of the 1950s embedded in the Cumberland Plan persisted. Ball (1996) observed, 'The one constant planning body during the life of the radial freeway proposals was the DMR, which by 1969 had developed into a highly independent bureaucracy and had repulsed all challenges to its authority'. The SPA published its *Sydney Region Outline Plan 1970–2000* (SROP) in 1968. The SROP retained the radial expressway system of the Cumberland Plan, though transport proposals were 'sparse' suggesting the greater influence of authorities such as the Water Board which cooperated with the SPA. The lack of necessary data in the SROP was corrected six years after its publication by the Sydney Area Transportation Study, 1974 (SATS), commissioned by the Ministry of Transport. SATS was based on the well tried US model of road transport already promulgated in Melbourne in 1969 and later in Perth in 1970 (Myer, 1993: 214). SATS proposed an integrated plan of highway development and public transport systems designed to serve projected land uses and meet anticipated travel demand. As with the 1969 Melbourne Transportation Plan, however, the proposals were weighted towards road infrastructure. The projected highway system incorporated 555 kilometres of freeways and expressways and 241 kilometres of new major regional roads. The freeway plans were opposed by the Whitlam federal government (1972–1975), and in September 1974 the Commonwealth *Roads Grants Act 1974* cut federal funding for urban arterials by over 50 per cent. The confrontations between the DMR and the inner city residents were backed by the Labor Federal Government and the Labor State Opposition (see Burgmann and Burgmann, 1998) and ultimately led to the demise of proposals for surface level inner urban motorways. Nevertheless, elements of the SATS proposals were retained by the RTA, even up to the present day.

A series of metropolitan plans for Sydney continued to appear from the State's planning agency. Embedded in the plan *Sydney into Its Third Century: Metropolitan Strategy for the Sydney Region* was the DMR *Roads 2000* plan which proposed an arterial road grid at around 3–4 kilometres spacing – including the current and proposed freeway network, cross-regional bus routes (assisted by the road network), and growth in road freight transport (with the onset of containerization). While the plan reserved no new corridors for heavy rail, the existing public transport system would be supplemented and upgraded with completion of the East Hills-Glenfield railway line, the duplication of the Blacktown-Riverstone line and a rail link from Merrylands to Harris Par (NSW Department of Planning, 1988: 52).

A small but significant shift away from roads dominance occurred in 1993 with the publication of a discussion paper (*Sydney's Future: A discussion paper on planning the greater metropolitan region*) to elicit public comment for new directions on Sydney's future growth strategies. The paper addressed the transport issue by posing two challenges: 1) the problems associated with travel to work and the need to reduce the distance people travel from their homes to work; and, 2) the need to increase the proportion of people using public transport.

The 1995 plans, *Integrated Transport Strategy*, and *Cities for the 21st Century* brought 'integration', both of land use and transport, and of transport modes into the discourse. But integration at the operational, as opposed to the aspirational, level was not evident. Toon and Falk (2003: 140) noted that the plan continued to talk about transport corridors without specifying what modes of transport were to be included. Continuing the theme of 'integration', a further plan, *Action for Transport 2010 – An Integrated Transport Plan for Sydney* came out in 1999. There followed a period of 'integrative' transport planning which in some ways foreshadowed that of Melbourne in the 2000s, and like that of Melbourne (thus far) was not able to operationalize the idea of integration. Meanwhile, throughout the late 1980s and 1990s outer metropolitan motorways were built under private-public partnerships: the M4 Motorway, an urban expressway of 40 kilometres, the M5 South Western Motorway, the M2 Motorway (a BOOT project), and the Eastern Distributor completed in late 1999. Government funded projects included the Gore Hill Freeway, the City West Link, Homebush Drive and the M7, connecting the M5 from the south-west, the M4, and the M2 to the north-west, thereby completing the metropolitan ring road. The 2000s saw the addition of several new inner city motorway tunnels: Lane Cove and the Cross City tunnel planned by the road agency, now the Roads and Traffic Authority. The outcome of transport planning over time in Sydney can be seen in Figure 7.2 below.

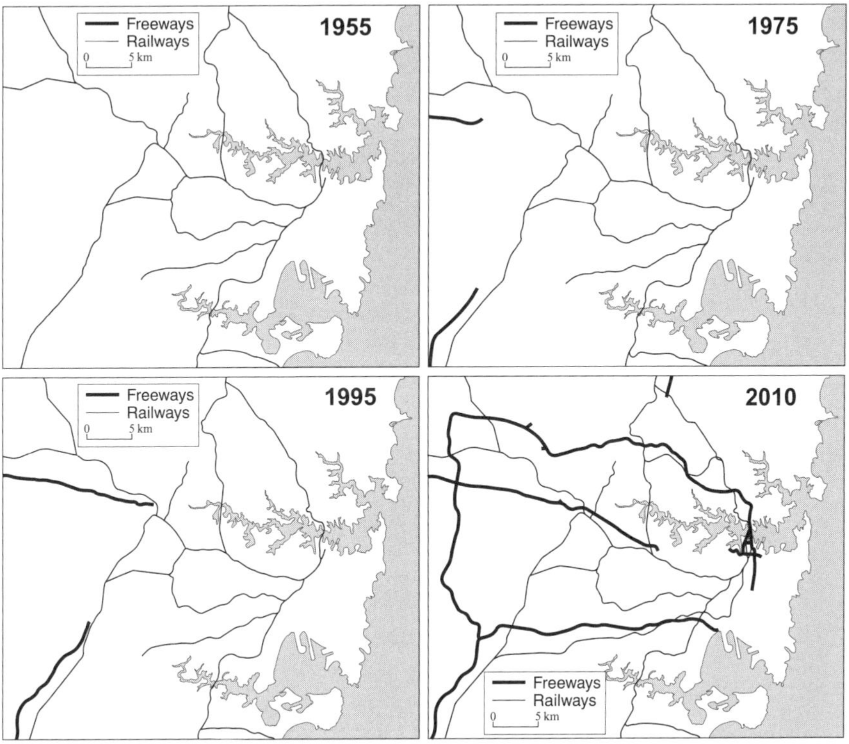

Figure 7.2 Sydney: Railways and freeways 1955–2010

Following the rather tentative steps towards integrative planning in *Action for Transport 2010*, the next metropolitan planning report published in 2005, engaged fully with the concept of integration, and it remains the basis of city planning up to the present (2010): *Metropolitan Strategy: City of Cities – a Plan for Sydney's Future*. The *City of Cities* plan identified five urban sub-regions and five major activity centres: North and South Sydney (the latter being the Sydney CBD), Parramatta, Liverpool and Penrith as well as a variety of smaller centres. The plan promised to expand the transport network to connect these places: 'investment in the rail network will provide faster and direct public transport linking suburban centres and areas where jobs are concentrated' (Government of New South Wales, 2005: 9). The plan is aspirational, emphasizing aims and visions for 'global Sydney' and its region, to accommodate population and employment growth and enable the city to compete on the world stage with other, implicitly successful 'global cities'.[4] But it included a range of specific transport projects. New investment in road infrastructure proposed by the RTA was to continue in the main outer urban growth areas: AU$2.3 billion in the South West and AU$1.3 billion in the North West. Other road infrastructure 'potential' proposals were included, 'subject to funding availability' (Government of NSW, 2005: 172).

There would be 'an investigation' of the need for road corridor reservations. But the major focus of the *City of Cities* plan was improving public transport. A 'Metropolitan Rail Expansion Plan' had been announced in June 2005 including a new heavy rail line from Chatswood to North Sydney and then under Sydney Harbour to the CBD (the 'Next Harbour Crossing'). A new North West–CBD–South West Rail Link was promised costing AU$8 billion, 'the largest public transport infrastructure projects ever seen in Australia' (ibid.: 165). This was to be a 'V' shaped (a 'V' on its side) line connecting major areas of population growth (ibid.: 176). The plan also promised a network of 'strategic bus corridors' with fast, frequent and direct services. Integrated ticketing for all public transport in Sydney was (again) promised, with improvements to transport interchanges. A new emphasis on investment in walking and cycling was introduced, acknowledging that one in five trips are made by walking or cycling, and that 'short trips currently made by car could also be made by walking or cycling, improving local accessibility, neighbourhood amenity and community health' (ibid.: 181).

The *City of Cities* plan embraced the concept of managing demand for travel and marked the resurgence of public transport infrastructure planning. However the further development of integrative planning in Sydney was attended by increasing conflict within the NSW government over public transport infrastructure projects. A variety of public transport projects were planned, only to be cancelled shortly thereafter. Glazebrook (2010) records the outcome of the

4 The report displays a preoccupation with ranking, Sydney being ranked as a 'beta' world city with San Francisco, Toronto and Zürich: a truly strange order in which Beijing, Shanghai, Washington and Berlin appear further down the list in the 'gamma' category. No other Australian city gets a mention.

internal political wrangling within the NSW (Table 7.3). He observes, 'There was a major "war" between those promoting a more compact city supported by public transport, particularly rail, and those strenuously opposed to any more expansion of the heavy rail network' (ibid.: 3). It is noticeable that a different Premier and different sets of Ministers smile out from the Forewords of successive reports. Between 2004 and 2010 NSW went through four Labor Premiers without a change of government.

Table 7.3 Key urban public transport projects cancelled or delayed in Sydney

Project	First announced	Original completion date	Fate	Subsequent developments
Integrated ticketing	1996	2000	Contract terminated 2008	New process restarted in 2008
CBD light rail	1996	EIS and other studies completed 1996–2003	Not proceeded with	Re-announced 2010
Bondi Beach rail link	1996	2002	Cancelled 1999	
Parramatta-Epping rail link	1998	2006	Cancelled 2003	
Hurstville-Strathfield rail line	1998	2006	Not proceeded with	
Liverpool 'Y' rail link	1998	2010	Not proceeded with	
North West rail link	1998	2010 (to Castle Hill)	Delayed several times and cancelled 2008	
South West rail link	2005	2017	Stage 2 deferred 2008	Stage 2 re-announced 2009
Next Harbour Crossing	2005	2017	Cancelled 2008	Parts re-announced in Feb. 2010
North West Metro	2008	2017	Cancelled Oct. 2008	
CBD Metro	2008 (Oct.)	2016	Indefinitely deferred Feb. 2010	
West Metro	Mooted 2009	Not defined	Indefinitely deferred Feb. 2010	

Source: Glazebrook, G. (2010) 'Generating Solutions for Sustainable Transport – the Sydney Experience' Table 1s. Paper given at the GAMUT International Conference, *Sustainable Transport in the Asia-Indo-Pacific, Varied Contexts, Common Aims* 2 to 4 June.

The integrative phase brought greatly increased commitment to new public transport projects and, following two major train crashes with loss of life, a concern for rail safety. But in Sydney more than in Melbourne or Perth rail projects were the subject of fierce political battles within the government. These battles in part reflected conflict within the bureaucracy, with Treasury and the RTA sceptical of the value of investment in new rail lines, whilst a succession of Premiers responding to public pressure with the help of the public transport agencies installed a succession of new rail projects in the plan for Sydney. What was plainly not at all integrated was the political process of the NSW Labor Party.

While the roads agency has been able to insert its plans and projects single-mindedly into the multifunctional and integrative plans from the 1950s to the present, the public transport agencies have made major contributions only in the last 10 years or so under political pressure for better public transport. They have, however, been inhibited by the lack of integration of public transport modes at the operational level (paralleled by the organizational fragmentation discussed above), and constant political instability responding to changes in the wider political agenda. Such was the public frustration at planning for public transport that rival plans were produced. In 2009 an independent inquiry was chaired by a distinguished transport planner, Ron Christie, a former co-ordinator general of rail, head of the Olympic Roads and Transport Authority (which had delivered successful transport outcomes for the 2000 Olympics), and Director of the RTA (see Christie, 2010). A key recommendation of this inquiry was for the establishment of a Zürich style public transport authority to be called 'Transport for Sydney' (Recommendation Gov 1).[5]

Integration, when operationalized, would mean that the separate companies operating public transport modes do not design the system of routes and timetables, the authority is the 'customer' of the operating companies and the primary face of public transport to the users. Operational integration means: integrated fare structures, electronic ticketing across all modes, planning of timetables to ensure seamless intermodal transfers, interchanges designed to a high standard, and planning processes 'which consistently seek to optimize the entire network rather than the interest of a particular operating company or mode of transport' (Christie, 2010, para 2.4.1).

Perth: Steps to Towards Operationalizing Integration

The first strategic plan for the metropolitan region prepared by Stephenson and Hepburn in 1955 proposed a compact region (see Table 7.4). Although two new suburban rail lines were proposed, a substantial regional road network focusing on radial routes from the Perth CBD was the mainstay of this multi-focus planning

5 The Liberal-National party coalition instituted 'Transport for NSW' (including city and rural areas) in 2011 to co-ordinate the State's transport services. It remains to be seen if and how this will be done.

period. Planning and building of an arterial road network continued to dominate transport planning up until the 1980s, with the 1970 *'Corridor Plan'* 'setting in concrete' the plans proposed by the first comprehensive regional transport study for Perth (PERT) in 1971.

Up until 1983 public transport was being seen largely as a job for buses. However a turning point occurred in 1983 with the reopening of the Perth to Fremantle railway, commencing a period of critical juncture in which extensive new rail lines were built – a development very different from what happened in Melbourne and Sydney (to be more fully explored in Chapter 9).

Table 7.4 Key transport and land use plans for the Perth metropolitan region

Year	Plan	Authority and representation
1955	*Plan For The Metropolitan Region: Perth And Fremantle*	G. Stephenson (ex- Ministry of Town and Country Planning, Britain, invited by the State government) and Hepburn (appointed as Town Planning Commissioner)
1962	*Metropolitan Region Scheme Report*	Town Planning Department
1967	*Rationalization of Urban Public Transport*	Director General for Transport
1970	*The Corridor Plan*	Metropolitan Region Planning Committee
1971	*Perth Regional Transport Study*	Steering Committee: Chairman and Commissioner Main Roads Department (MRD); DG Transport; Town Planning Commissioner; Deputy Under Treasurer; Study Director (MRD). Study Team: MRD (7 members); Town Planning Dept. (3); Maunsell & Partners (1)
1978	*Development of the Public Transport System in the Perth Metropolitan Region*	Director General of Transport
1979	*Urban Public Transport for Perth: Rail and Bus Policy*	Ministry of Transport
1982	*Transport 2000: A Perth Study*	Director General of Transport; Department of Transport (DoT)
1988	*Development Control Policy 1.6 Development around Metropolitan Railway Stations*	Department for Planning and Urban Development

continued....

Year	Plan	Authority and representation
1990	*Metroplan: a planning strategy for the PMR*	Department Planning and Urban Development (DPUD)
1991	*Road Reserves Review Final Report*	DPUD, DoT, MRD and Transperth
1993	*Metropolitan Transport Strategy–Perth: A More Responsive and Integrated Public Transport Service for Perth.*	Transperth and Department of Transport WA
1995	*Metropolitan Transport Strategy*	Department of Transport, Ministry for Planning, Main Roads Western Australia (1996) Metropolitan Transport Strategy
1996	*State Transport Policy*	DoT
1997	*State Planning Strategy*	Western Australian Planning Commission (MfP)
1997	*Liveable Neighbourhourhoods: Community Design Code*	Ministry for Planning /Western Australian Planning Commission
1998	*Better Public Transport: Ten Year Plan for Transperth 1998–2007*	Department for Transport
2004	*Network City: Community Planning Strategy for Perth and Peel*	Department for Planning and Infrastructure/ Western Australian Planning Commission
2009	*Directions 2031: A Draft Spatial Framework for Perth and Peel*	Department for Planning and Infrastructure/ Western Australian Planning Commission

During the late 1960s and 1970s Perth saw a period marked by planning focussing on a ‘bus only’ strategy (*Rationalization of Urban Public Transport*, 1967). Several possible avenues for ‘rationalization’ of suburban public transport were proposed, meaning: ‘the reforming or re-arranging of the public transport system so that its inefficient parts are made to function better or if necessary are excised altogether’ (DGT 1967: 1). While some suggestions were made for increasing the efficiency of the bus operations, the main focus was on rail services, and it was these that needed to be ‘excised’. This argument was based on two main premises: first, that rail services made a disproportionate contribution to the public transport ‘deficit’ (patronage was declining so expenditure per passenger was high compared to buses); second, that Perth did not have the population density to justify a rail system.

The 1979 *Urban Public Transport for Perth: Rail and Bus Policy* (Ministry of Transport WA, 1979) continued to argue the government’s public transport policy platform. The decision had been taken to close the Perth-Fremantle rail line, and therefore the main function of the document was to explain the factors influencing the decision to close the line which included: declining patronage and no potential for growth, capital cost savings, operating cost savings, efficiency in energy usage, road planning benefits, removal of railway freight traffic (ibid.: 4).

In the 1978 policy document, *Development of the Public Transport System in the Perth Metropolitan Region* the argument for more federal roads funding was strong – there was 'growing inadequacy of some parts of the road network' which, it was claimed, resulted from 'a paucity of road funds'. With this context established, the Department of Transport stated that the Government's intention for the public transport system was first, to continue to develop the bus network where the regional road system is or will be adequate, and second to strengthen the ability of rail to contribute to the public transport demand in those corridors where it is unlikely, without massive expenditure on roads, that the bus network would be able to perform adequately (1978: 12).

A joint report by Transperth and The Department for Transport produced in 1993 marks the turning point for integrative planning. The report – *Metropolitan Transport Strategy, – Perth: A More Responsive and Integrated Public Transport Service for Perth* (Transperth and DoT, 1993) highlights the strong desire by the transport agencies for integrated transport planning. The strategy is critical of transport policies that do not price travel appropriately and, consequently, encourage urban sprawl in Perth. The report describes 'four ingredients of a proven recipe' to tackle the problem of car dependence and create a more public transport orientated city. It is argued that these ingredients should extend beyond simply providing public transport 'hardware' (infrastructure) to focus also on 'software' (service, reliability, information) and policies supportive of public transport (1993: 2). So rather than propose public infrastructure solutions the report proposes bus priority measures and car park pricing in the central city. It is in this report that the case for a new organizational structure was made, whereby the task of determining public policy was separated from transport service delivery, and local transport advisory groups were to be established to improve community liaison. The aim, under this new arrangement was to, 'develop and implement policies to achieve economic, social and environmental goals while the operators focus on transport efficiency and effectiveness' (ibid. 1993: 14). The difficulty in bringing forward the 1993 report's aspirations for public transport can be seen in the later publication of the *Metropolitan Transport Strategy* (MTS) (DoT et al., 1995). This was an interesting document since it was produced by the Department of Transport in consultation with the other transport and planning agencies and efforts were made to cement the agency relationship by placing all the agency names as authors. The strategy, however, was never endorsed by the Planning Commission and so sat outside formal planning implementation procedures. Nevertheless it marks an attempt (at least in published policy) to change professional culture, at least for some within the state agencies.

In 1998 the first 10-year plan for public transport was published, demonstrating the shaky commitment to planning for both buses and rail. Early in the plan it was made clear that the plan does not promote radical changes (bus ways are preferred to railways), but although the plan adopted the 'welfare' approach to public transport it also argued for promoting public transport as an alternative for car drivers. The importance of the public transport system for both users and non-users alike started to emerge. Several times it is argued in the document that more government resources need to be allocated to public transport, but that this funding commitment would only come if there is greater public pressure. The main focal point of the radial public transport system remained Perth's central city, but there was some shift toward promotion of 'inter-suburban' linkages and better connections amongst suburban 'activity centres'.

By the late 1980s the transport integration arguments develop further with an interest in integration with land use planning. The introduction in 1988 of a State 'Development Control Policy 1.6 Development around Metropolitan Railways' required that all planning applications on land adjacent to metropolitan railway stations support rail use and access by providing for higher density residential development. While a pioneering policy for integration through transit oriented development, its implementation has been slow to take effect (Curtis, 2011). Although this is a policy produced by the State planning agency, this agency's direct relationship to the State Planning Commission provides the platform for cross agency integration. In addition to the department – ministerial relationship, land use and transport planning are managed through a tripartite model that interconnects the Minister, State Planning and the independent West Australian Planning Commission (WAPC). The Commission, originally established by Perth's first planner Stephenson in 1955, is responsible under the Act for guiding land development in the State, coordinating aspects of the planning process and maintaining the partnership between the community, business and government. The Commission also coordinates the transport planning and land use functions through its two sub-committees – the Sustainable Transport Committee and the Infrastructure Coordinating Committee. These committees provide direction and decisions on all transport infrastructure (road and public transport), particularly where it requires reservation of land.

Later, other state government initiatives developed which supported this change in direction towards land use transport integration. The 1997 State Planning Strategy emphasized the inadequacies of physical and social infrastructure on Perth's suburban fringe. Of great significance, however, was the introduction in 1997 of a state-wide planning code, *Liveable Neighbourhoods*, promoting the development of 'walkable communities' where day-to-day activities can be served within a local area (WAPC, 1997). The LN Code was commissioned by the WAPC from two teams of consultants. Perth-based Taylor Burrell had been responsible for much innovative residential design in a series of new urbanist-influenced brownfield projects in Perth, and Melbourne based Ecologically Sustainable Design (ESD) Pty Ltd had developed a charrette-based approach to neighbourhood design applying the Responsive Environments principles developed by Ian Bentley (1985) and his colleagues in the UK, but with a stronger ecological dimension (Morris & Kaufman, 1997). The consultants worked with a Steering Committee of ten State politicians, planners, transport engineers and development industry representatives who oversaw the development of the Code. The 'movement network' occupies nearly half of the planning code with requirements to encourage walking and cycling and disabled access; and the provision of public transport. A set of traffic management guidelines emphasized interconnected streets and their detailed design and landscaping (WAPC, 2000), and it is here that the first real moves to operationalize integrated solutions for all modes of transport were achieved. A small urban design unit was established within the Ministry for Planning Urban Design and Major Places (UDMP) to promote and implement the Code on greenfield sites, but it quickly became involved in a wider range of projects including developments around proposed new stations on Perth's expanding commuter rail network, retrofitting a 1980s car-based regional shopping centre (Mirrabooka), expanding a dynamic local centre adjacent to a rail station (Claremont), restructuring an underdeveloped transit-served regional centre (Armadale) and redeveloping high amenity brownfield sites (Leighton Beach and South Beach). All of these projects have now been implemented. The outcome (Figure 7.3 below) for road and rail infrastructure in Perth shows how the city has invested in new railways as well as new roads.

So far the above strategies and development control policies had seen a change in transport and planning practices among some agencies and in some areas across the city. A gap remained in that these strategies were operating in a vacuum at the strategic/metropolitan level. In 2004, after an unprecedented process of public deliberation, the Network City (WAPC, 2004) metropolitan planning strategy was published and strongly articulated the paradigm shift towards integrative planning that was emerging in a small way in the earlier strategies. Network City's proposed a spatial framework designed to achieve the integration of land use and public transport networks by focusing on accessibility planning (see Curtis, 2008 for details). While providing a broad planning framework it became clear that implementation would require significant changes to traditional planning tools, practices and professional skills. The WAPC recognized this need in adopting nine priority tasks representing the work needed to achieve the paradigm shift (Table 7.5). The work required for tasks 1–6 was clearly interrelated; it was not possible to deal with each task sequentially. To deliver this would require a change of existing work practices, requiring a number of inter-disciplinary project teams working in an integrated way that allows for the cross-fertilization and development of ideas working across the state planning, transport and infrastructure portfolios.

Some clear actions have been taken to deliver the new paradigm. A cross-agency committee was established, tasked with identifying station precincts where public interventions (both land use and integrated transport) could be made (Mouritz and Ainsworth, 2009). This has resulted in small scale transport infrastructure changes but land use change has been much more difficult to achieve. The fastest progress has been where the redevelopment model has been introduced; Subicao station

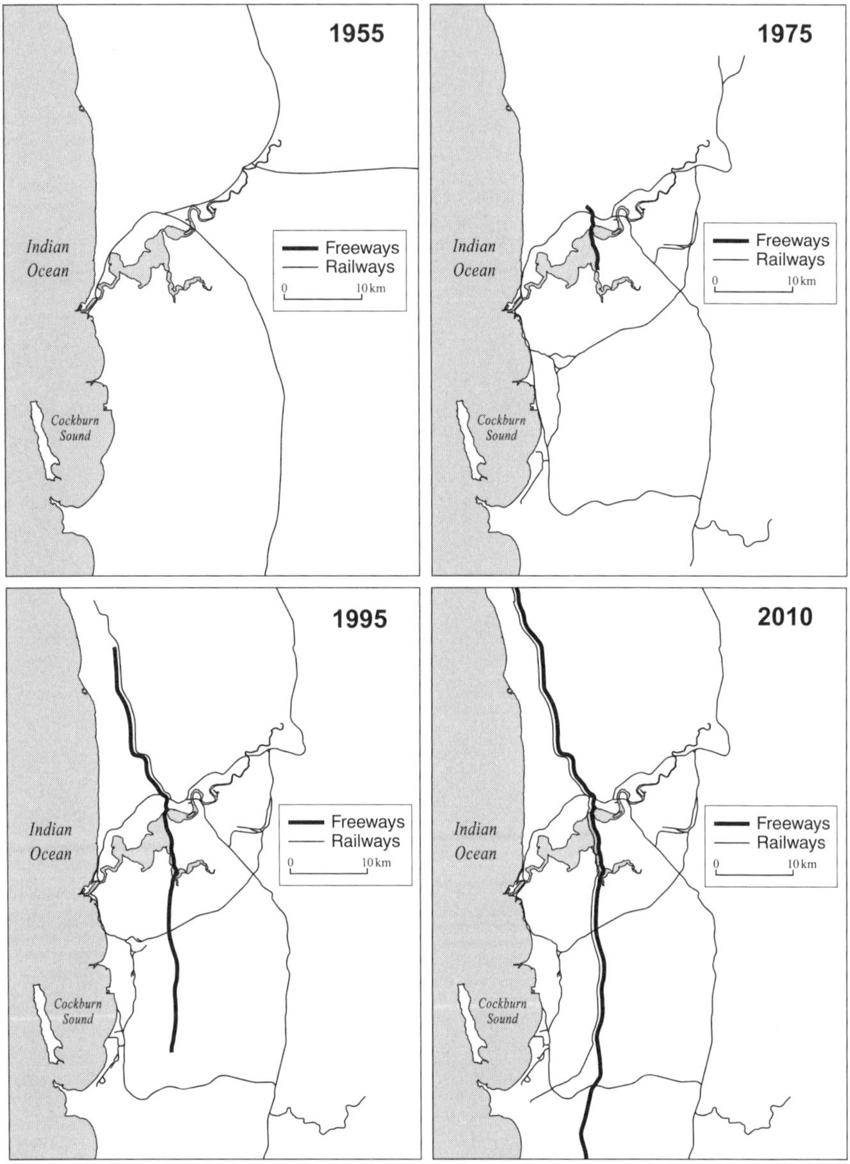

Figure 7.3 Perth: Railways and freeways 1955–2010

Table 7.5 Implementing 'Network City': WAPC's nine priority tasks

TASK	Involves...
1. Detailing the metropolitan structure	Expanding and explaining the overall metropolitan structure of activity centres, activity corridors and transport corridors (including the principles that will govern their local planning and management).
2. Population, housing and job targets	Determining and applying methodologies for setting these (and other targets) collaboratively.
3. Managing growth	Locating, staging and managing infrastructure and Greenfield and infill development.
4. Developing the activity centre concept	Determining the character, location and management of activity centres.
5. Developing the activity corridor concept	Determining accessibility principles as well as the character, location and management of activity corridors.
6. Developing the transport corridor concept	Determining the character, location and management of transport corridors.
7. Enhancing institutional structures and decision-making	Enhancing and improving the way all levels of government and civil society work together.
8. Relating sustainability to decision-making	Planning and managing growth according to sustainability principles
9. Accessing time, money and skills	Ensuring that all stakeholders have access to skills and funds needed to deliver planning through participatory mechanisms.

Source: WAPC, 2005.

precinct provides the best example of this (Howe, Glass and Curtis, 2009). The southern suburbs railway opened in December 2007 and includes 11 new railway stations in the first. During the construction phase there has been considerable development activity within the corridor, evidence that the railway has acted as a catalyst for new development. Indeed many property developers choose to promote access to the railway as a key draw card in their development (harking back to the past railway suburbs of the 1880s!).

Up until Network City, there had been no practice of considering accessibility to activity centres by public transport or in directly integrating service provision with land use planning. Instead the public transport agency aimed to meet a minimal service level primarily as a community service obligation rather than a real alternative mode of travel to the car. An urban structure capable of supporting an efficient public transport service suggests particular parameters for the composition, size and location of activity centres within the network of activity and transport corridors. Public transport providers and land use planners would need to work together to define this in relation to public transport capacity. Finally, in 2010, this kind of detailed planning work was embraced by the state agencies of planning and public transport. The 2009 metropolitan planning strategy, *Directions 2031*, has been produced by this mode of working, and, still to emerge from Cabinet at the time of writing, the next 10 year public transport plan has utilized this approach (see Curtis and Scheurer, 2010 for details).

Conclusion

A movement towards integrated transport planning is clearly apparent in all three cities to varying degrees, and with varying methods and emphases. 'Integration' can mean integrating public transport modes, integrating road planning with public transport planning – with particular emphasis on infrastructure, integrating active modes with public transport, integrating the planning of freight with passenger transport, and integrating planning of the whole transport system with land use planning. Perth has gone further than the other two cities in demonstrating land use transport integration in practice. Indeed one can say that in some ways Perth has moved from integrative to integrated planning.

It is also clear that the movement has a long way to go before real integration is achieved. Neo-liberalism, still the mainstream ideology of governance in Australia, opposes government planning, preferring to depend entirely on market incentives. The movement towards integration could in principle be reversed by the action of determined political leaders. But would such political leaders find support in the community, whose expectations of greatly improved transport systems have been aroused? There is enough in what we have found to suggest strongly that a critical juncture has been reached in which the path to integration will in fact be hard to reverse.

Chapter 8
The Discourse of Roads

In this chapter and the following three we turn to a different level of empirical research, looking at the discursive aspect of path dependence. This chapter focuses on the 'discourse of roads' as evidenced in our three case study cities: Melbourne, Sydney and Perth. We use the 'storyline' as a conceptual strategy for evaluating the argumentative content of documents in order to analyse the discourse surrounding planning to improve the road system. The first section, then, explains the storylines we used and how they were constructed. We then turn to an analysis of key policy documents produced since the 1950s by roads and urban planning agencies, which together can be termed the 'transport policy', for each city. Of key interest is whether, and if so how, sustainability gains traction within policy. By mapping storylines in transport policy it is possible to compare their relative strength over time and between the three cities. The final section examines the way in which these storylines support decision making for major road infrastructure projects in each city.

Constructing Storylines for Transport Policy

The storylines of transport institutions are the discourse of explanation and justification, the talk that goes on around policy, sometimes written, sometimes verbal: the storylines justifying and explaining public policy (Hajer, 1997). In a storyline, understandings and objective diagnoses are inseparably interwoven with values and norms. A storyline is essentially an idea which can be expressed as a claim, and can then be put together with other storylines to form explanations, justifications or arguments. Storylines function to hold discourse together across the range of actors in a particular policy domain. They may also serve to position actors relative to one another.

The storylines we found were first identified through analysis of planning documents used to support a proposal for the construction of the Melbourne Ring Road (which began life in 1929). The storylines which emerged were the product of a purely inductive investigation derived from these documents over an extended period of study. The research was expanded, first in 2002[1] looking at transport policy for roads in Melbourne and Sydney, and again in 2004[2] where metropolitan Perth

1 Research supported by grants from University of Melbourne and the University of Western Sydney in 2002.

2 Research supported by a grant from the Australian Research Council – Discovery Grant DP04500709.

was added, together with public transport policy for all three. At that point a more methodical approach to the investigation of storylines in transport and planning discourse ensued. This resulted in some refinement of the storylines to capture and compare the experience of all three cities and to draw on the authors' own practice experience. Our distinct purpose was not simply to identify the existence of these storylines in published policy, but also to discover whether these storylines were still believed by key players in the transport policy domain (see Chapter 11).

We grouped the storylines under four main discipline headings: engineering, economics, town planning and environment (Table 8.1). This was on the basis that the road building program is continually justified to politicians and the public by a network of storylines from three professional sources: engineering, economics and urban planning. Together they provide a common language uniting the various groups and interests that lobby for roads when the need arises.

The 'Engineering' storylines start with 'predict and provide', that is – predict future trends in transport mode use based on past travel behaviour and provide infrastructure to meet this projected demand. The storyline is strongly grounded in the methods of behavioural science. 'Predict and provide', then, assumes an ideal situation of 'free movement' in which individuals and firms are able to move wherever and whenever desired without hindrance or congestion. Congestion is therefore portrayed as bad (as of course it is for car users, but not necessarily so for cyclists and pedestrians where lower vehicle speeds improve the safety of these most vulnerable road users). The next storyline argues that there should be 'balance' in transport policy. Although this is appealing at face value, it was found that this is not an even balance applied but, rather, one weighted in favour of road transport. Sometimes, this weighted 'balance' is described in terms of 'integrated transport' in which the weighting towards roads is justified with reference to a variety of sub-storylines: the need to relieve bottlenecks, create by-passes and join up 'missing [road] links'. Finally, the engineering story is accompanied by the assumption (often implicit) that transport problems will be solved by future technological advances: non-polluting fuels, low-energy vehicles, advanced traffic management software, and so forth.

The 'Economic' storylines start with 'travel as opportunity', where travel is believed to provide people and firms with opportunities to improve their conditions of life and wealth, to which they have a right. The underlying assumption that travel is essentially a 'good' is connected with a series of assumptions: that the best way of providing people and firms with travel opportunities is by leaving the task to private corporations or citizens; that perfect competition to meet travel demand in a 'free' market is possible and desirable (including the minor storyline that car ownership and use represents a modern lifestyle); and that building transport infrastructure meets latent demand to travel, which, if met, adds to increased consumer satisfaction. There are two other basic assumptions: that the public interest (social welfare function) in transport means the aggregate of individual preferences expressed in the travel choices each individual makes; and, that infrastructure investment contributes to economic growth of the city as a whole. This last storyline supports the minor storyline that new road investment is necessary to provide for rapid transport of freight (although the road space is always used by cars). Other minor storylines include personal choice, including those concerned with symbolic values of car use: sex appeal, escape, exciting lifestyle, wealth, freedom – more than the convenience they provide.

The town planning storylines are underpinned by the economic and engineering rationales and are grounded in the idea that metropolitan planning strategies should always seek to reduce the travel time for individuals within the urban system. This finds expression in several spatial and social dimensions: areas of low population density cannot be served effectively by public transport (building roads is sometimes justified as providing for bus transport, but, unless there is strict segregation of traffic in favour of buses, the roads also serve to increase private vehicle transport); an argument for locational equity: that low-density outer urban areas, which are poorly served by public transport, must be compensated by investment in new and improved roads; that improved mobility for the many may be traded off against reduced amenity for the few whose homes are close to new road infrastructure projects. Minor storylines include: that the continuing vitality of the Central Business District depends on good road connectivity with the suburbs; and so it follows that suburban development should be concentrated into corridors along transport spines (growth corridors); better roads are needed to link households with locations of concentrated employment (growth poles); and freeways can continue to be built because the land has been reserved, regardless of how long ago the reservation was made.

The storyline for 'Environment' argues that an ecologically sustainable city cannot be achieved while governments continue to favour investment in road transport for private car use over alternative modes (public transport, walking and cycling). Moreover, scarcity of fossil fuel will in future make it necessary for cities to adapt their transport systems to make the best use of alternatives to private car transport. Freedom of movement for all people and freight in cities cannot be achieved without severe increases in traffic congestion, loss of lifestyle and environmental values. While it is possible to break the environment storyline down into sub-categories (such as pollution, fuel shortage, emissions trading etc.) we have not done so, instead preferring to distinguish it clearly as the oppositional storyline. As counter arguments to other storylines it is argued that economic benefits can be achieved by increasing accessibility rather than providing for maximum mobility, together with those storylines on the costs of travel: costs to life and health, social costs of 'car dependency', the total financial costs to the public of a road-based transport system including a hidden government 'roads deficit' (Laird and Newman, 2001; Tranter, 2010).

There are a number of miscellaneous related storylines: that fixed rail systems are obsolete and buses are better (linking to ideas about the modern city); that public transport provision is a welfare solution rather than offering a real alternative transport mode (and that car use is inevitable so we should not cater for public transport as an alternative); that road construction is a response to 'public demand' or 'community pressure'; that the only viable transport mode choice for most people in Australian cities is use of the private car (car dependency).

Table 8.1 Storylines

Storyline	Argument
Major storylines	
Engineering	
1) Predict and provide (science)	Discover the trend in the use of different modes of transport and project it forward a given number of years. Then design a transport system to fit the projection. This story fits well with 'separate markets'. It tends to portray opposing forces as 'unscientific', and probably 'emotional' as opposed to 'rational'.
2) Chronic congestion versus free movement (disease)	An assumed ideal situation where the whole population (individuals and firms) of a city is able to move wherever and whenever desired without hindrance or congestion. Congestion suggests an *abnormal* condition (Concise Oxford Dictionary) – a disease, some threatening disturbance of a bodily system. Because it is 'abnormal' the storyline also suggests a *normal* condition which in this case is free-flowing traffic. This is a powerful story because of the high reputation of medicine and the fears associated with human disease, and because it bears a message which cannot be disputed because it is unspoken.
3) Balanced transport (doublethink)	Ostensibly this storyline suggests investment and/or infrastructure provision in equal balance (implicit in the word itself)' between different modes of transport, but the words 'equal' or 'even' balance are carefully avoided (or worse, actors argue "we weighed all modes but there wasn't room for say on-road cycle lanes or separate bus-lanes as well as through traffic lanes, right and left turning lanes and parking lanes"). Balanced transport therefore *appears* to mean equal balance but *in fact* means balance in favour of roads. It is hard to oppose 'balance' because the opposite is 'imbalance' but this is an example of doublespeak. It is also linked to arguments about road safety (but in fact this means for vehicle users primarily rather than cyclists or pedestrians).
a) Bypass and bottleneck	*'Bottlenecks' occur where road capacity in the system is uneven and restricted. The road capacity must therefore be increased. New roads may be built to 'by-pass' congested areas. The advantage of these stories is that they justify a series of partial increases in road building without apparent change to the transport system as a whole.*
b) Future technology ('technofix')	*road infrastructure must accommodate new technologies. New technologies justify road infrastructure investment. Technofix – technology will solve any current problems with roads/cars*
c) Missing links	*ideally a city should have a uniformly good road system throughout the urban area in which there should be no 'missing links' in the road network.*
d) Safety	*it is important to invest in new road building to avoid major safety issues on existing roads.*

Economics	
4) Travel as opportunity (justice)	All travel results from people seeking opportunities to enhance their lives by engaging in spatially separated activities. Travel is therefore a 'good', a means to a pleasurable end, a benefit rather than a cost. It is connected with the idea of mobility as a right and a freedom. It is an argument grounded in human reality. Opportunities *are* spatially separated. People *do* experience spatial separation and seek to overcome it. They *do* benefit individually thereby. They *do* yearn for face-to-face contact. Goods *do* have to be distributed physically to people. This, however, is not telling the whole story since it overlooks the occasions when the same or greater benefit can be achieved without travel or with less travel, where time can be saved by working/shopping from home, where time can be wasted by unnecessary travel, where time and money can be saved by rationalizing the logistics of delivery.
5) Social rationality from the aggregate of individual decisions (economics)	If individuals are confronted with the full price of their travel options, with travel seen as a scarce commodity, they will make a rational choice of mode of transport. The aggregate of individual rational choices results in a socially rational choice. The storyline is used to justify a system prioritizing individualized transport and road pricing as a solution to congestion.
6) Infrastructure investment causes economic growth (economics)	Investment in road infrastructure causes increase in GDP and savings (travel time) in costs to business (and individuals) and therefore stimulates economic growth. Linked also to the storyline above.
7) Separate markets (economics)	each mode of transport has its own market which it serves. The market consists of existing users of different modes not the consumers of transport as a whole. This storyline is backed by the institutional reality that each mode of transport is provided by separate organizations.
8) Public transport deficit	The dominant storyline about public transport since the 1950s has been that public transport is running a 'deficit', that is the difference between necessary operational expenditure and the revenue derived from fares and other minor sources. This deficit has to be covered by a 'subsidy' out of the State Budget. The contribution that public transport makes to reducing traffic congestion on the roads is not counted in this calculation.
9) Public transport equals welfare (inevitability)	public transport (being obsolete and second rate) is provided for people who cannot afford to run a car, and is therefore 'welfare transport', a residual mode.
e) Modern lifestyle, way of life, the right to drive	*Extreme version of 4 above – cars = progress (fits with 'rail obsolete' storyline below)*

Storyline	Argument
f) Flexibility/ individualism	*individuals plan their own timetables*
g) The people's choice	*people have chosen cars for mobility*
h) Expert versus citizen	*linked to above – the rhetoric seems to have shifted from community vs. expert in the early 1980s to increasing community participation as time went on.*
i) Freight (also part of 6)	*roads are not really for cars but for increases in freight transport – we must keep freight moving – Linked to storylines about congestion and economic growth.*
Town planning	
10) Low urban density can only be served by roads (planning)	Public transport cannot efficiently serve areas of the city of low gross density.
11) Locational equity (social democracy)	The balance of costs and benefits for residents varies in different locations in the city. People should not suffer disadvantage because of where they live. Used in conjunction with 8 above to justify road building in outer suburbs.
12) Balance between mobility and amenity	The argument runs as follows: (i) mobility is necessary for modern life; (ii) mobility can be best achieved by use of the private car (the latter is assumed but rarely stated); (iii) but cars decrease amenity (e.g. pollution, noise, severance); so (iv) the best resolution of the inherent conflict between mobility and amenity is to restrict most traffic movement to arterial roads, which will preserve amenity on all other (residential) streets.
j) CBD vitality	*The prosperity of the CBD must be protected by ensuring good road based connectivity with the suburbs.*
k) Urban corridors	*Growth of population and housing should be directed into linear corridors along a transport spine.*

l) Growth poles ('significant existing employment locations')	*It is possible to detect existing areas of the city that can be described as key regions for economic performance, and potential future performance.*
m) Inner freeways bad, outer good	*Freeways can be built in places where there are existing road reservations for them.*
Environment	
13) Sustainability (environment)	The major storyline in opposition to roads investment in outer suburbs. The current transport technology and modal mix is imposing unacceptable costs in public health, local pollution and global climate change.
Other minor storylines:	
n) Fixed rail systems are obsolete/ buses are best	*Fixed rail transport systems (train and tram) were developed in the nineteenth century and have been superseded by a superior form of mass transport, the private car. Where you must have public transport, buses, which use roads, are more flexible and cheaper. Originally part of an idea about modern cities.*
o) Community pressure	*Road capacity increases are a response to political pressure from 'the community'.*
p) Car dependency	*In the past it has been used to support the 'rail obsolete' and 'separate market' storylines. The city is car dependent therefore we must invest in the things cars run on: roads. Now – Australian cities are unnecessarily dependent on car transport and roads.*

Note: Major storylines are numbered 1–13 and highlight in grey; Minor storylines are shown in italic font and 'listed' alphabetically; Most (but not all) of the master storylines are mutually reinforcing.

Having described the storylines, we turn to the published 'transport policy' of our cities to ascertain the presence, or otherwise, of each storyline and how they change over time and between each city.

Transport Policy – The Discourse of Roads

Conventionally, policy documents are portrayed as explanations of a government's intentions and actions. As Pierson (2000) states, policies, grounded in law and backed by the power of the state, signal to actors what has to be done and what cannot be done. Although policy documents are published and seen as a response to the need for public accountability, generally the lay public rarely reads them. Rather, policy documents are scrutinized by the network of actors with particular interests in the policy in question. From this perspective, policy documents have the function of a cultural record of the discourse network, reinforcing dominant belief systems.

Policy texts themselves are open to multiple interpretations, and are influenced by the content and interpretation of other policy texts. Policy interpretation and emphasis can also change with the political climate in which policy is interpreted (Ball, 1993). It is, nevertheless, important to gain an impression of the way in which policy texts are presented in published documents, since these policies represent the public face of the particular agency. For our three case study cities these published policies provide a historical record of the storylines prevalent in the road and public transport policy fields since the 1950s. We took the 1950s as a starting point since by this time all three cities had completed a first metropolitan planning strategy, it also marked the period of significant urban expansion which has literally cemented in place today's urban morphology. This provided an appropriate launching point from which to explore the way in which transport policy was constructed.

Selecting policy documents proved a challenge since there was no one document that could be pointed to as 'the transport policy' prevalent at a particular period in time. Instead we made our selection using a two-step process. First we searched State library and public transport/planning agency library catalogues in order to assemble a comprehensive list of transport and land use planning policies published since the 1950s. Second, we consulted key local experts (academics, long standing planning practitioners, politicians and public officers) in each city asking them to indicate which of the full list of policy documents they considered to be representative and influential in shaping city development and transport infrastructure in the city. A list of the final selected documents is provided in Appendix 1.

Each policy document was analysed using the storyline framework above. Our approach was to identify the presence, or otherwise, of the storylines and to identify the presence or absence of evidence used to support these storylines. We used a systematic approach, collating the evidence using a proforma. This enabled us to record the number of occasions a storyline was used, extract the quote and link it to any supporting evidence. We analysed written text and visual texts –

photographs, graphics and pictures. While this form of content analysis is clearly a qualitative research method, we found it helpful to map the number of occurrences for each storyline as a means of comparing the emphasis between storylines as well as change over time as well as comparing across the three case study cities.

Melbourne

The 1954 *Melbourne Metropolitan Planning Scheme* (MMBW, 1954) starts our period of analysis and we conclude appropriately at 2000 with the *Challenge for Melbourne: Issues in Metropolitan Planning for the 21st Century* (DoI, 2000). Two things are immediately evident from the analysis of storylines over time (Figure 8.1): first that the 'Engineer's Tale' has maintained a presence throughout the 50-year period; second that the 'Environment' storyline has been present for a considerable part of the period since the mid-1970s. The 'Economist's Tale' is seen to reinforce the 'Engineer's Tale' from the early 1970s, reinforced also by the 'Town Planner's Tale'.

The 1954 Plan described a comprehensive system of radial routes, radiating outwards from an inner ring road. The objective was to select routes along which controlled access would be possible for the greatest distances in order to reduce travel time, an acknowledgement of the problem with the spread of the metropolitan area even at that time. Three photographs (MMBW, 1954: 100) showed 8 and 6 lane arterial roads and were captioned 'Modern road practice – American examples of urban roads'. 'Predict and Provide' clearly underpinned the strategy, and the solution of 'balanced transport' was determined by providing high-speed arterial roads first and public transport, possibly, second:

> When a system of high-speed arterial roads has been established, the most rapid transport between the central area and some of the large residential areas could probably be provided by buses running express from city terminals to the edge of the areas concerned...The alternative of frequent services of feeder buses connecting with railway services at suitable stations would result in overall travel times from 5 to 10 minutes more than express bus routes. (ibid.: 193–194)

In fact the storyline 'fixed rail obsolete/buses best' underpinned the transport strategy, where it was argued that fixed rail is seen as a worldwide problem with the growth of motor transport. The 'people's choice' storyline was also used to justify the approach, seen as 'placing greater emphasis on those forms of transport which prove most effective under present-day conditions' (ibid.: 167), and 'lifestyle', 'So far as our planning is concerned, it is not unreasonable to plan for the time when practically every family will own a car and commercial vehicles will have increased in like proportion' (ibid.: 167).

'Predict and provide' continued as the master storyline in the 1969 *Melbourne Transportation Study* (MTC, 1969): the objective was 'to provide a transportation service matched to the expected demand' (ibid.: 3). The storyline has the persuasive advantage of appearing scientific because it depends on quantitative methods, but the internal logic of the transport models was never revealed. This study, produced by a State advisory body (The Metropolitan Transportation Committee), was a three volume study of Melbourne's transport system prepared under the technical direction of a US consultant, Wilbur Smith (together with a local consultant: Len T. Frazer and Associates, Melbourne). The mode of choice was private car, where the plan was for motorists to be able to travel on freeways at an average speed of not less than 35 mph in peak traffic (ibid.: 4) compared to 30 mph by rail and this was seen a 'proper balance between private and public transport' (ibid.: 1). The benefits of a system of freeways were seen in avoiding 'considerable congestion' (ibid.: 50), and 'failure to provide road space would have a serious impact on commercial goods movement and the cost of transporting goods by road' (ibid.: 50) – thus the roads solution was supported by the 'Economist's Tale'. In justification of the roads solution we see the minor storylines of 'rational planning' and 'expert vs. citizen', where the reader is informed of how many weeks work of professional engineers and economists it took to test the options using computer modelling.

Roads planning was supported in Melbourne in early land use and transport strategies, and this discourse was further strengthened during the 1970s by the 'Town Planner's Tale'. The 1971 Plan (MMBW, 1971) indicated that Melbourne should be encouraged to follow a corridor form with urban development confined to the 'growth corridors', separated from each other by 'green wedges' of open country protected from urban development' (ibid.: 4) – thus we see the urban structure moulded to fit the earlier transport strategy. Even where planners viewed some residents, correctly, as transport disadvantaged, the response continued to be one of road building throughout the 1980s. The 1980 Metropolitan Strategy (MMBW, 1980) noted that activities were increasingly dispersed and so public transport was needed to be provided in a dispersed way (an argument for a bus-based system). By 1983 it was stated that 'public transport is to cater for those who do not have use of a car' (MoT, 1983: 25) added to which was the view that 'for many trips and many situations public transport is not suitable' (ibid.: 30). So by 1987 it was noted: 'Urban sprawl has made it difficult to provide an efficient system of public transport in outer areas' (DPE, 1987: 10).

A longstanding concern of planning has been to protect amenity. However, when it has to be acknowledged that building new transport infrastructure destroys local residential amenity, planners aligned with the engineer, arguing that a balance must be sought between the 'values' of mobility and amenity. The argument ran as follows: mobility is necessary for modern life and can best be achieved by the use of the private car. But cars reduce amenity (pollution, noise, severance of communities). So the best resolution is to restrict most traffic movement to arterial roads which will keep traffic out of residential areas and preserve the amenity of residential streets. It was presented in detail in the report of the Melbourne Metropolitan Board of Works in 1981 but it has been referred to since when justifying expenditure on arterial roads or 'by-passes' (Lonie, 1980; MoT, 1986). The storyline appealed to the rhetoric of pragmatic compromise, which is trying to do the best by most people, most of the time. It evoked a utilitarian argument: for the greatest happiness of the greatest number a few small groups may have to lose out: 'The effects of road improvements are such that a small group [those living in the vicinity of arterials] may suffer while the majority [those living in residential streets] benefit' (MoT 1986: 22).

By the 1980s 'predict and provide' had begun to be modified or masked by other storylines. The 'Economist's Tale' was strong in, for example, the 1980 Lonie Report to the Victorian State Government: 'mobility has become an ingrained part of the 1980 lifestyle' (Lonie, 1980 para 3.10); and 'the general public has evolved its own transportation system, which is basically a private enterprise one involving specific and personal investments by millions of individuals in mobility' (ibid.: para 11.1). This storyline was also associated with the view that private vehicles are the modern way to travel, a storyline that maintained traction right up to 1996:

> Travel by car offers considerable personal advantages to the user. It is convenient, available at call, offers a door-to-door service in most instances, provides privacy and security in a personal space, and is widely affordable. The access to social, economic (especially work) and recreational destinations which the car provides has become an essential component of the lives of most people. The availability of the car has enabled people to live at their preferred locations in suburban or rural environments while at the same time being able to participate in a full range of urban activities. (Victoria, 1996: 62)

Transport economists mostly assume that travel itself is a cost to the individual measurable in time taken. There are, however, also costs to society and to the environment which are less easy to quantify. In the comparison between cars and public transport, the freedom to spend time doing something while travelling other than driving a vehicle (reading a book or newspaper, writing on the laptop, sleeping, eating and chatting) is never considered in these policy documents. In conventional analyses, this individual cost is mitigated in two ways: if the opportunity thereby accessed outweighs the cost (as is claimed in benefit-cost analysis), and if individual control over the journey is maximized. Two important points about this storyline are, first, that it positions those who oppose road building as seeking to take away freedom of movement and self-expression; second that it supports an important related story that investment in infrastructure brings economic growth.

The economic justification for road building continued to be reinforced during the 1980s. The major planning policy document of the Victorian government elected in 1982 states:

> Cross-metropolitan travel is difficult in Melbourne ... Most traffic uses the radial routes going through the most congested areas of the central city. Also, movement of freight through urban areas creates serious problems – the economic need to provide rapid access for freight traffic is emphasised. Establishing an outer

> ring route would overcome many of these pressures and provide better linkages between key parts of the metropolis. (DPE, 1987: 10)

Finally an economic storyline that stems from transport practice in Australia and shores up 'predict and provide', is the 'separate markets' storyline: 'The rich mosaic of overlapping land uses, historical patterns of development, redevelopment opportunities, and growth areas leads to a multitude of distinct travel markets, each with its own needs and requirements' (DoI, 1996: 13). In this storyline each mode of transport has its own market which it serves. The possibility of one mode substituting for another is denied (for instance better public transport relieving road congestion). This storyline is backed by the institutional reality that each mode of transport is provided by separate organizations (see Chapter 5).

The emergence of the 'Sustainability Tale' appeared as early as 1974, in a report on objections to the 1971 Plan (MMBW, 1974). We see sustainability expressed in terms of rising petrol prices, limited resources and concern for the environment. These are themes revisited in the 1979 review of development strategies (MMWB, 1979), again in 1989 (MPE, 1989), and in 1992 (DPH, 1992). Despite this caution the 'Economist's Tale' won out, with the 1978 Transport Plan arguing for freedom of choice (State Co-ordination Council, 1978: 12) – 'Travel as Opportunity', and the 'difficulty of providing public transport especially during off-peak periods' (ibid.: 25). The Engineer's tale also won out, arguing for the need to bypass bottlenecks (ibid.: 6).

The rhetoric of sustainability emerged more strongly in the discourse of the 1996 policy documents. *Transporting Melbourne* linked a discussion of 'environmentally sustainable transport' to the European report on urban travel and sustainable development (ECMT, 1995) and concluded that Melbourne was probably well placed to capitalize on the European ideas[3] (DoI, 1996: 13, Inset Box). The report went on to state that 'environmental sustainability is essential not only for its own intrinsic value, but also because it is a key ingredient in making Melbourne an attractive place to live and work, which is essential to its long term economic development' (ibid.: 15). The policy document, however, did not introduce innovative European solutions for transport, instead it entrenched the plan for Melbourne's orbital freeway which would be seen later to encourage more dispersed urban growth and increased use of motor vehicles:

> Many cities enjoy the benefits of a ring or orbital corridor, and Melbourne – with its low density and large geographic area – is the sort of metropolis likely to benefit most. To be most effective, a Metropolitan Orbital Transport Corridor needs to be a continuous corridor fully circumscribing the metropolitan area. (DoI, 1996: 32)

---

3 These include adoption of best practice in land use and transport planning to reduce the rate of growth in car travel; integrated transport management, including congestion pricing, parking control, public transport priorities, investment in transport efficiency; and international cooperation to reduce dependence on fossil fuels.

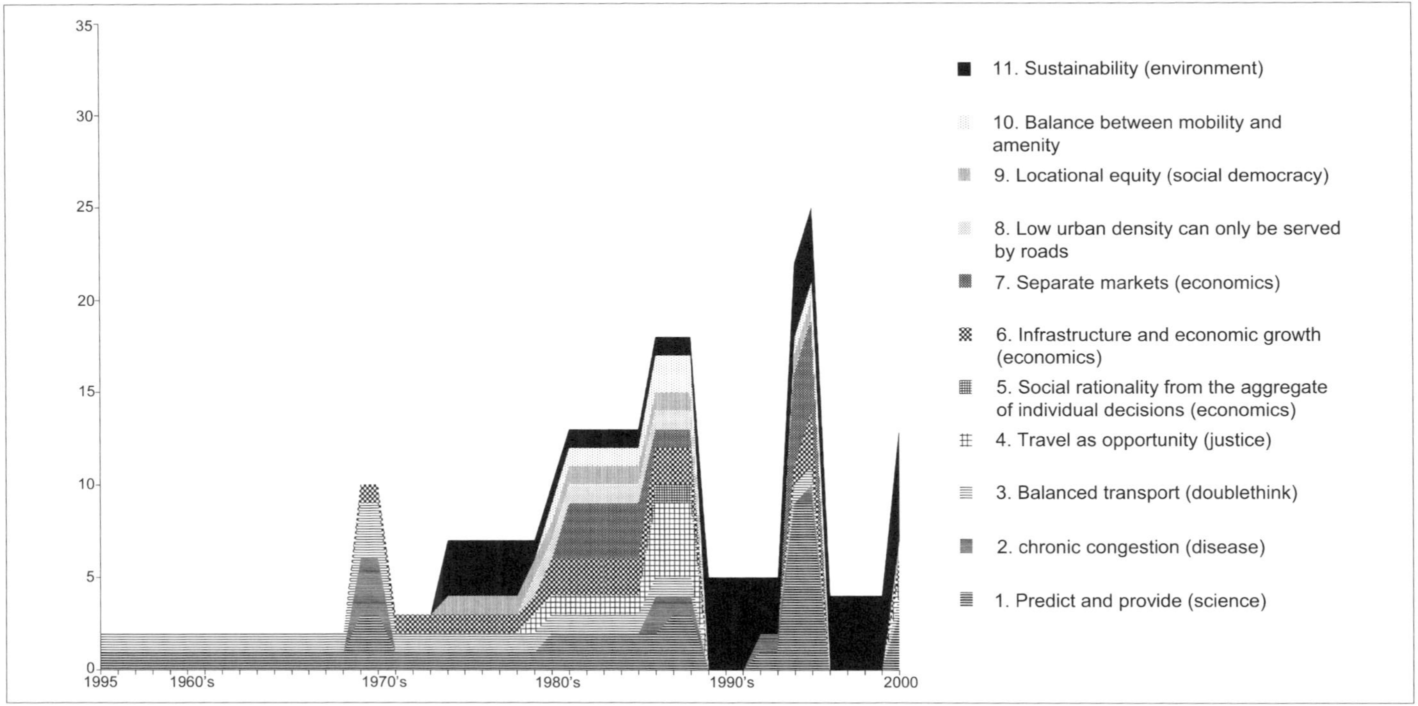

Figure 8.1 Melbourne: Roads policy discourse – Storylines over time

Note: Eleven major storylines are shown – style of hatching depicts storylines as grouped by the four 'tales'. Chart shows number of occurrences of storyline in policy documents over the period that policy applied. Each storyline is stacked one on top of the other – the height of the storyline depicts the number of occurrences, for example during the mid-1990s there were four occurrences of the sustainability storyline in policy documents.

Two final policy documents were published at the end of century; both were strong on sustainability messages. The *State Planning Agenda* (Thwaites, 1999), the first planning document put out by a new State Labor Government, stated: 'The Government is committed to building the principles of ecologically sustainable development into the process of decision making and program delivery across the whole of government' (ibid.: 20). As before, though, this was tempered by the need for 'balance': 'The Government will ensure that decision making on the use and development of land reflects a sensible balance between economic development, social growth and cohesion, and the sustainability of Victoria's environment' (ibid.: 20). Car dependency was placed on the agenda, with town planning and urban designers charged with resolving this, rather than transport planners: 'the Government will ensure the adoption of less car-dependent urban design' (ibid.: 23). Finally in 2000 the rhetoric of sustainability was strengthened (DoI, 2000). Arguments were put for development to be concentrated around public transport nodes (ibid.: 21), and it is noted that urban fringe development occurs in areas with little public transport (ibid.: 29). It was also acknowledged that planning for roads and rail had not been equal: 'Melbourne has often established corridors for future road projects, but has less often reserved land for rail' (ibid.: 34).

Sydney

We start in the 1960s and analyse 14 policy documents culminating towards the end of the century with two policy documents: from town planning the 1998 *Shaping our Cities* (NSWDURP, 1998), and from transport planning the 1998 *Action for Transport 2010* (NSW, 1998). As with Melbourne, Sydney's 'Sustainability Tale' emerged in the mid-1970s (Figure 8.2) but never really gained traction against powerful debates around 'balanced solutions'. The 1960s was dominated by the 'Engineer's Tale' which, like Melbourne, continued to the end of our period. The 'Town Planner's Tale' has also been strong, reinforcing the 'Engineer's Tale' by favouring dispersal of activity to avoid congestion in the CBD.

In 1967 the State Planning Authority published its *Prelude to a Plan for Sydney* (SPA, 1967) which noted a forecast doubling of population growth coupled with rising standards of living. The concern was how to cater for forecast congestion ('predict and provide') given the 'desire to own and enjoy the convenience of a motor vehicle' (ibid.: 51), part of the Economist's Tale – minor storyline 'the right to drive'. The solution was seen thus – 'much of the future planning of Sydney will be concerned with the increased facilities needed – roads for vehicles on the move, parking space for vehicles at rest' (ibid.: 51). True, the document did acknowledge the need to provide for public transport ('balanced transport') with the focus on rail and peak traffic and the concerns of the bottle-neck of access into the Sydney CBD via the Harbour Bridge. The solution, though, was 'a new major urban highway (the Western Distributor) …

between the wharves and the central business area' (ibid.: 56) in order to keep freight moving and separate it from city centre traffic. Public transport was seen as an intractable problem both because it was seen as operating at a loss and because people preferred private transport.

The 1968 *Sydney Region Outline Plan* (SPA, 1968) cemented this approach. Its main concern was avoiding over-concentration of employment in the CBD, a continuation of earlier concerns from the 1940s. As such, the single most dominant storyline in the document was the Engineer's one of 'chronic congestion'. It was argued that this has 'been severely accentuated by the use of the motor car as a mode of transport to work' (ibid.: 12), but from this, the conclusion was drawn that car trips need to be spread over a wider area, rather than reduced. The suggestion of discouraging car travel to the central city in favour of public transport was not made. Instead, it was accepted that it is the 'age of the motor vehicle' (ibid.: 27, 28) necessitating 'a highway grid aimed at reducing the dominance of the radial pattern focussed on the metropolitan city centre and at increasing the accessibility of the coastal areas (ibid.: 15). However, the first fold-out map at the back of the document (entitled 'Sydney Region Outline Plan, 1970–2000 A.D.') showed radial expressways which almost without exception lead into the centre. It is only the penultimate map, the 'Sydney Region Principles Diagram (Central Coast) 1970–2000 A.D.' that showed a grid, with two north-south 'primary roads' to the west of the city centre. Of interest is the first map which included a legend differentiating between 'railways existing' and 'railways proposed', but 'expressways' were all represented by the same symbol, suggesting perhaps that these were non-negotiable. The plan recommended that road funding be increased by 600% (from $14.5 million to $87 million per annum) to cover up to 350 miles of new freeways which may be required to provide an efficient main roads system (ibid.: 101).

The *Sydney Area Transport Study* (SATS, 1974) was the eventual completion of the transportation study recommended by the State Planning Authority in 1968. Public transport was considered but weighted by the 'balanced transport' storyline: 'The overall philosophy of the Study has been to attempt to develop a balanced transportation system for the Sydney Region' (ibid.: VI-13). Here, 'balanced in the sense of combining the best use of motor vehicles with the best use of public transport vehicles … an appropriate balance aimed at maximising public benefit and minimising public cost' (ibid.: XIII-1). However their recommendations for expenditure saw 60% for private transport (a new grid of highways, other road improvements, parking facilities and traffic management) and 40% for public transport. Despite the rhetoric of 'balanced transport', SATS was still firmly embedded in the 'predict and provide' regime: 'It is recommended that future transportation plans allow for strong growth in motor car ownership and vehicle miles of travel. Failure to do so will restrict the opportunities for travel by motor car drivers in the Sydney Region' (ibid.: VIII-12).

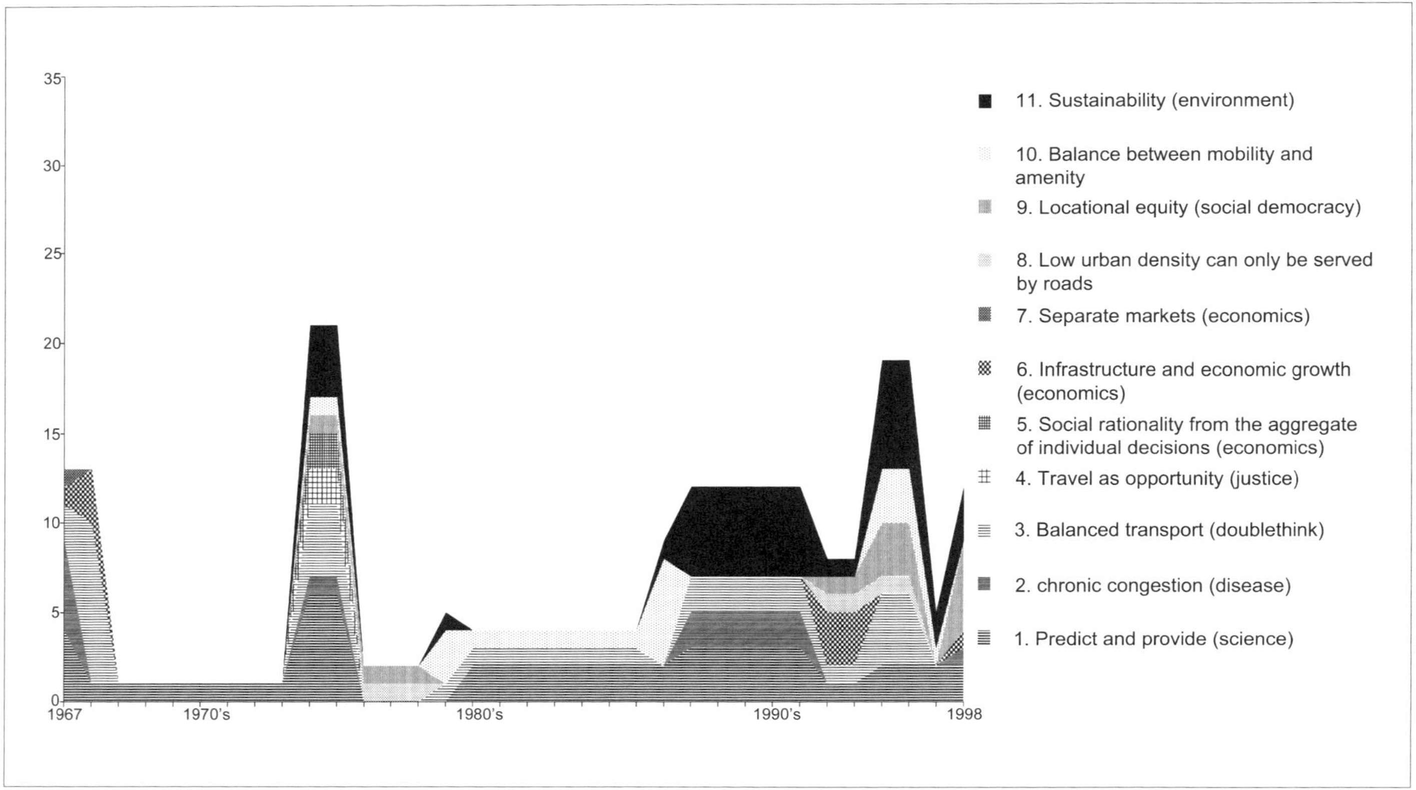

Figure 8.2 Sydney: Roads policy discourse – storylines over time

By the mid-1970s (following an anti-freeway community revolt), through the deliberations of the Urban Transport Advisory Committee, we see the first suggestions to reconsider freeway planning as Sydney faced a more financially constrained transport future: 'Apart from the questioning of the efficacy of freeways by the community, in particular those directly affected by their construction and operation, transport planners both in Australia and overseas are now beginning to reappraise long-term freeway and other transport plans' (URTAC, 1976: 38–39). By 1979, however, the argument was for funding to be increased for both roads and public transport, interestingly recommending half as much funding again for public transport as for roads in the period to 1979–1991 (URTAC, 1979: 32–33, 45). The Committee's core suggestion for funding this increase was for the Main Roads Act 1924 to be amended 'to enhance the flexibility of current funding arrangements' (ibid.: 9).

At the start of the 1980s the 'predict and provide' approach continued to be questioned: 'In the past the emphasis has been on provision of road space to meet projected traffic growth under conditions of relatively free flow and low operating costs. In future, the capacity of the transport networks must be taken substantially as given … Land use planning goes hand-in-hand with measures designed to allocate priorities for use of road space, in the process of adaptation to a constrained transport future' (PEC, 1980: 39). However, by the end of the decade the discourse of the Department of Main Road's *Roads 2000* returned to 'predict and provide' with proposals based on the old 'predict and provide' assumptions (DMR, 1987: 6) and the assurance that that this was an 'affordable plan', including only 'essential freeway links' (ibid.: 5), an assurance which was unsubstantiated because no cost figures were given in the document.

The unresolved tension between 'predict and provide' and 'sustainability' continued throughout the 1980s and 1990s as the ecological sustainability tale emerged in the discourse only to be modified again and again. The planning document *Sydney into its Third Century* provided a plethora of reasons why improved arterial roads are important: they allow employers to locate away from the public transport system; they maintain and increase accessibility; they take traffic off local streets; they reduce congestion, improve road safety and cater for increasing car ownership; and perhaps most importantly, they facilitate road-based public transport and freight (DEP, 1988: 16–20). The need to reduce trips by private vehicle was accommodated by the implicit suggestion that transport-land use integration (ibid.) combined with public transport improvements (ibid.) would be be adequate to minimise use of the private car. In 1993 a discussion paper on planning Sydney's future (DoP, 1993) acknowledged the value of public transport in reducing congestion and emissions but also recommended road improvements in order to keep freight moving and because public transport could not effectively serve areas of low gross density: 'residents are forced to rely on private transport' (DoP, 1993; 1995: 14 and 16). Again the sustainability discourse faltered. Ideas for travel demand management were discussed, including road pricing, but by 1995 *the Integrated Transport Strategy for the Greater Metropolitan Region* (DoP, 1995b) in noting that 'pricing' could help to allocate transport resources more efficiently, only discussed pricing for public transport. Discussion on pricing for road users was conspicuous by its absence (ibid.: 20). Moreover, although environmental consciousness was a feature, it was undermined by the discourse that the economy is ultimately more fundamental to human welfare than the environment: 'Efficient freight and commercial transport is fundamental to regional, state and national economic welfare, although environmental effects can be significant' (ibid.: vi).

As the end of the century loomed the failure of the 'Sustainability Tale' to gain real traction is evident. *Cities for the 21st Century* teetered between viewing the understanding of environment as fundamental (DoP, 1995a: 3) and the understanding of environment as merely one among other goals (ibid.: 15), the latter being susceptible to 'trade-offs' (ibid.: 112). The inherent conflict between improving roads and reducing car dependency remained unresolved – the intention was clearly to direct 'employee and client travel' on to public transport, leaving arterial roads for freight and commercial goods transport' (ibid.: 55) but the measures for doing this remained unresolved. In 1997 the road solution was again justified with reference to the needs of freight and public transport (RSPGMR, 1997), this despite the problems for the environment: 'The planning and development of new strategic road links now also recognizes the potential implications their operation can have for increased individual car travel and air pollution as well as land use patterns which are contrary to the principles of a compact city'. However the only solution suggested was the provision of public transport along the new roads (ibid.: 51). *Action for Transport 2010* continued with an acknowledgement that 'car dependency' must be reduced (NSW Govt, 1998: 14) side by side with the claim that roads must be improved for freight (ibid., 1998: 28–29) and road-based public transport (ibid., 1998: 30).

Perth

We start in 1955 with Perth's first metropolitan planning strategy and complete the analysis in the late-1990s with the State Planning Strategy. From a list of 59 potential policy documents, 13 documents were selected and analysed. This selection was based on a review by local experts who considered these the key influential policies, plans and reports. As with Melbourne and Sydney, the 'Engineer's Tale' dominated the transport policy discourse. While 'predict and provide' supported the discourse throughout the period, the 'balanced transport' storyline emerged as a key component of the 'Engineer's Tale' in the late-1960s, and strengthened during the 1980s. The 'Economist's Tale' provided a support role through the whole period, emerging in the mid-1960s as a discourse around infrastructure and economic growth. From the 1950s through to the 1990s this discourse had been underpinned by the 'Town Planner's Tale with the storyline 'low density city can only be served by roads'. The 'Sustainability Tale' can be seen occurring twice: as a brief glimmer in the 1970s and again in the final year.

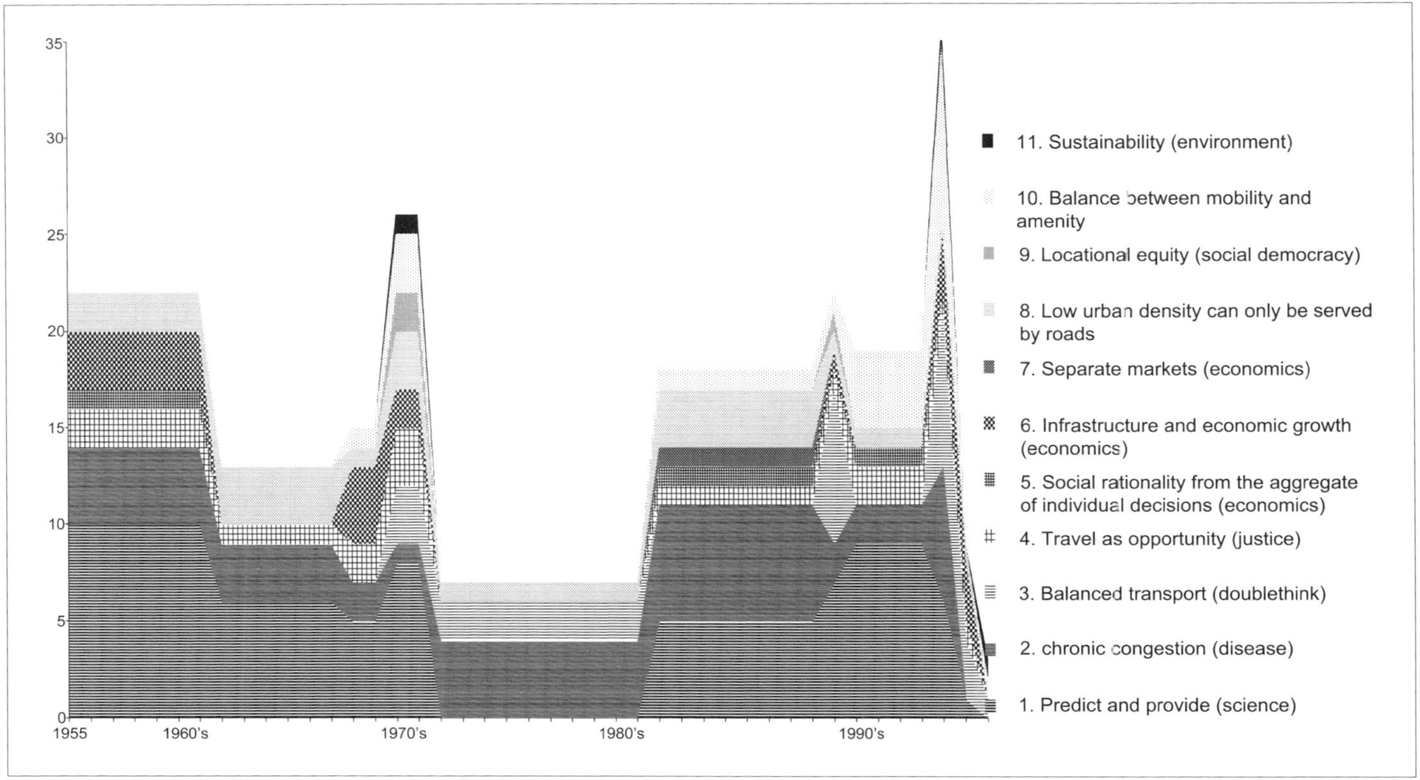

Figure 8.3 Perth: Roads policy discourse – storylines over time

The Stephenson-Hepburn plan (1955) exemplified the Master Planning approach predominant in the 1950s and 1960s. The plan employed authoritative language, using science and the argument of economic efficiency to substantiate assertions and proposals. The storyline of the plan was heavily grounded in the 'science' of predict and provide: 'It becomes increasingly evident that the growth of motor traffic will soon justify the provision of motorways … if road transit is to be accomplished without unnecessary waste of time or undue risk of accident' (ibid.: 110). As well as a 'scientific' base, the discourse employed emotive language: 'An increase in the volume of traffic cannot be escaped, and without early provision for it – including the necessary resumption of land – the chaos and congestion of the City's roads and streets in the near future may not be beyond rectification' (ibid.: 176). Coupled with 'predict and provide', the plan was highly focussed on roads for economic prosperity (efficiency): 'The economic loss caused by an inadequate road and transport system is considerable – a fact that is recognised by all responsible road authorities' (ibid.: 106).

In 1962 the Metropolitan Region Town Planning Authority re-examined the 1955 report to consider its continued validity. The Authority deviated little in its approach, arguing that 'private vehicles therefore will continue to predominate as long as it is possible to use them for the daily movement to and from work (Town Planning Department, 1962: 20) and 'from the knowledge of traffic growth, the alarming degree of accidents on sub-standard roads, and for other reasons, it is evident that a road framework of much higher standard and greater magnitude than contemplated in the 1955 Report has become necessary' (ibid.: 20).

A car-based transport solution was also supported by the 'Town Planner's Tale': 'a further characteristic of the Region which militates against the expansion of public transport services is the low density of urban and suburban development' (ibid.: 20). By the end of this decade the need for: 'substantially improved and new more efficient highways' (MRPA, 1968: 34) did not abate. Now supported by the expectation of Perth as a modern city, comparing Perth with other cities (unnamed) around the world: 'Studies in almost any urban area anywhere in the world show a surprising degree of similarity, in that they indicate that there is a need of approximately one mile of freeway to every 10, 000 to 12, 000 inhabitants, and some two miles of major arterial streets to every mile of freeway. … it should also be borne in mind that at present some 55 miles of freeway should already be in use, instead of which we only have some four miles constructed' (ibid.: 38). The 'travel as opportunity' storyline was pervasive: 'the car owner wants to be able to move swiftly around the region and to be able to park his car as closely as possible to his destination …' (ibid.: 6), even for discretionary travel: 'the Sunday drive accompanied by massive congestion will no longer be regarded as pleasurable and there will be pressure to spend more on recreational amenities and, just as importantly, the mobility and freedom to reach them' (ibid.: 40).

The 1970s marked an important period in Perth's transport history. The *Perth Regional Transport Study* (1970) contained the argument that Perth was a growing metropolis which required a modern transport system to service the future needs of the region. This provided the basis from which a road-based transport network for a growing Region was founded. The Plan continued the reliance on 'predict and provide', but we also see the emergence of 'balanced transport': 'a balanced transport system might be developed...balanced is used in the sense of matching road and public transport with the needs of the travelling public' (MRPA, 1970: II-8). However, the alternatives evaluated in the report were all based on removal of the rail system and, instead, providing bus-based transit along freeways. 'Balanced' in this sense did not mean to provide equality in mode choice and availability.

The transport network proposed in the transport study was then reinforced by the urban development plan, the *Corridor Plan for Perth* (MRPA, 1970). The Plan was predicated upon corridor growth and development of sub-regional centres to provide employment areas within the corridors. The report stated: 'the Plan recognises the vital importance of movement within the area, and the interaction of traffic and land use. Consequently, close contact has been maintained with the work of the Perth Regional Transport Study...' (ibid.: 6). The *Regional Transport Study* and the *Corridor Plan* both came to the same conclusion, that linear growth corridors would provide 'the most economic transport system for the region' (ibid.: 6). The Corridor Plan established that there was no intention by town planners to reanalyse the transport issues, stating that the 'plan comments only on the general principles of transport as far as they are needed to understand the proposals made by the plan' (ibid.: 7) – referring here to the *Perth Regional Transport Study*. The pattern of growth proposed by the Plan was given credence by the *Perth Regional Transport Study* having undertaken research into urban form. It concluded on the same development pattern as the Corridor Plan. What is interesting about this particular period is that it was also a time when the 'Sustainability Tale' emerged, although it is clear that it was not a winning discourse: '... it [the transport system] seeks to preserve the environment but at the same time to provide adequate means of improvement ...', the environment was undefined, but one assumes it referred to the natural Perth environment, essentially a lifestyle preference, as environmental issues were not at the forefront of public concern at that time.

By the 1980s, the report *Transport 2000* (DGT, 1982) established the preferred strategic direction for the Office of the Director General of Transport and continued the technocratic and quasi-scientific approach to planning. The introduction to the report noted that 'no major transport study these days is complete without its computer to analyse masses of information, undertake complex calculations ...' (ibid.: 4) and: 'Large quantities of numerical information have been used in conjunction with mathematical techniques to simulate the overall travel behaviour of the residents of Perth' (ibid.: 57).

While Transport 2000 continued the 'predict and provide' approach, arguing for road provision to cope with forecast traffic congestion, the emergence of a more balanced view than in previous reports was evident. The need for public transport was acknowledged and methods discussed to improve the patronage for transit, 'By the turn of the century, it is likely that over 40 per cent of the people who work in the Central City will be travelling in by public transport ... This should lead to a major relief of traffic congestion pressures which would prevail if today's trends were to continue' (ibid., 125). Although public transport was seen as an alternative solution, once again low density urban form was the argument used against it, 'No public transport system could come near to meeting all the needs for journeys to and from work, let alone the multitude of other travel needs. The system which will come closest must be very flexible, and able to travel on suburban streets' (ibid.: 60). Reinforcing the approach is the minor storyline, 'lifestyle':

> It may be a faintly amusing concept to many of us to imagine Perth people, crammed together in a transport-efficient city on the edge of the wheatbelt and outback, and next to the void of the Indian Ocean. Our climate promotes the 'quarter acre' and 'fifth-acre' block: we seek space for garages and gardens, pools and patios, barbeques and boats. Our fences proudly proclaim our individual private spaces' [ibid.: 65]. In fact, transit supporters within the community are depicted as an alternative minority 'a few Perth residents are prejudiced in favour of public transport, probably because they see it as being environmentally superior. (ibid.: 52)

Where *Transport 2000* failed to turn towards public transport, by the 1990s a more tempered approach was evident in *MetroPlan* (DPUD, 1990), 'Metroplan recognises the importance of planning for the private car, but seeks to strike a new balance between private and public transport' (ibid.: 24). The recognition was that public transport would be important for the journey to work, 'particularly to and from the Perth Central Area' (ibid.: 9). The plan set a target for infill development of 20% in an attempt to stem the spread of the city:

> Perth has acquired a well-deserved reputation as a city that has learned to live with, and plan for, the motor car. So far, it has managed to reap the benefits of high levels of personal mobility without excessively suffering from the traffic congestion, urban blight, pollution and loss of amenity problems that other cities around the world have faced. This is partly a result of careful planning but, it is also related to the size of the city. (ibid.: 62)

Transit oriented development was promoted (ibid.: 23), with the first signs that land use could be planned to support public transport.

It was not a complete change. 'Balanced transport' remained a strong storyline, 'retaining the character of Perth, which has made the city an attractive place for so many years, must be a priority for future planning' (ibid.: 5). Road planning was to continue, seen in the emergence of the 'locational equity' storyline, 'there will still be a need to provide a number of major new roads if residents in new communities are to have comparable access around the metropolitan region to people in urbanised areas' (ibid.: 25). The plan proposed extensions to the main north-south freeway, completion of the orbital freeway and a comprehensive road network in new urban areas, albeit tempered by reference to 'public transport spines' in development corridors. (ibid.: 63). While predict and provide was relentless, occasionally the outcome was different, 'measures will be taken to reduce and remove existing long-standing road widening proposals where these are no longer necessary to meet forecast demands' (ibid.: 26).

In 1995 the first policy document[4] that explicitly recognized some of the impacts from excessive levels of car use was published – Metropolitan *Transport Strategy* (DoT et al., 1995) – 'under current trends the total metropolitan transport task, in terms of vehicle kilometres, would almost triple between 1991 and 2029, an increasing percentage of that transport would be undertaken using single occupant cars. This would have dramatic implications for road congestion and traffic delays, air quality, fuel use, accidents and injuries and costs' (ibid.: 10). In spite of this, and the mention of sustainability, the primary response was to seek greater efficiency from the road network, 'Car drivers will continue to enjoy good roads and will benefit from measures taken to promote high vehicle occupancy rates and greater use of alternative transport options … businesses will continue to have access to a good road transport system, enhanced by the development of a road hierarchy including measures to limit the impact of traffic congestion and delays' (ibid.: 15).

Towards the end of the century Perth policy makers continued to struggle with the sustainability storyline. Two final policy documents highlight the institutional tension. The *State Transport Policy* (DoT, 1996) continued with the 'balanced transport' storyline, 'The road network will be maintained and extended where necessary within a balanced transport framework' (ibid.: 5); and again linked with 'predict and provide' – 'Central to the strategy is the need to find a balance between meeting the growing demand for car travel and promoting other more efficient and sustainable transport modes'(ibid.: 14). But now the freight storyline was used to provide the rationale for continued road building, 'The efficiency of the transport system for the movement of goods and provision of services is vital to the Perth economy. Priority will be given to facilitating the movement of service and commercial vehicles. Access for these vehicles to industrial areas, transport terminals and interchanges will be improved' (ibid.: 17). As with Sydney, no strategy was put forward as to how to provide for the small proportion of freight movements without also providing enhanced access to private car drivers. In contrast we clearly see the sustainability mantra:

> The protection, enhancement, and repair of our environment and the sustainable use of our resources will determine our standard of living and place in the world in the 21st century. The aim is to ensure that the wealth to be created in future years will provide a significant impetus to deliver the necessary environmental improvements without reducing the wealth of future generations. (ibid.: 8)

4 Two years earlier a report – Metropolitan Transport Strategy-Perth: A More Responsive and Integrated Public Transport Service for Perth – was produced by Transperth and Department of Transport.

In relation to transport, again 'balanced transport' was regarded as the way forward, 'Perth will be an efficient city where the less mobile are able to easily access facilities and where there is a balance between walking, cycling, public transport, car and truck usage' (ibid., 55), although later roads were still viewed as being needed for freight (ibid.: 60).

The differences in the roads discourse of the three cities are rather small, despite an important factual difference: Perth has invested heavily in new public transport infrastructure and has initiated a degree of operational integration in the co-ordinated timetables of buses and trains on the new lines. This did not happen in Sydney or Melbourne during our time period, though later events show Sydney and Melbourne beginning to follow the Perth model. What is most striking in all three cities is the emerging contradiction between the sustainability storyline and others that were deployed to justify road building. This contradiction, and the way it was papered over becomes clearer when we examine case studies of orbital motorways in the three cities.

The Case for Roads Infrastructure – Case Studies of Orbital Motorways

In order to deepen our investigation into the discourse in transport policy we now turn to three case studies of road infrastructure projects in order to assess the extent to which the storylines found in published transport policy are applied in practice in the decision documents used. Each city has seen the development of an orbital motorway, following the pattern set in many cities, for example the M25 in London and the Peripherique in Paris. Given the considerable length of time that orbital freeways take to come to final fruition, our case studies focus on the decision debate relating to one 'link' of the motorway. The main published document used for analysis was the environmental impact statement.

Melbourne: Scoresby Freeway

The Scoresby Freeway (now called East Link and no longer 'free') is the tolled eastern section of a ring road around metropolitan Melbourne. The main argument for the motorway is contained in the Environment Effects Statement (EES) (DoI, et al., 1998a). Four options were considered ranging from: Option 1, the 'base case' – including only the road and public transport improvements previously planned, to Option 4 (construction of the full length motorway including some public transport improvements) (DoI et al., 1998a: S-5 – S-11). The consultants assessed each option using a set of criteria: transport, prosperity, liveability and sustainability. The main storylines found to support the full motorway option recommended by the consultants can be placed within the Engineer's Tale ('improved transport performance) and the Economist's Tale ('increased economic prosperity'). Of the two other criteria – 'improved liveability' was only weakly supported by the consultants' analysis, and 'improved sustainability' was not supported. Interestingly the 'improved liveability' criteria included the measures 'enhance regional accessibility' and 'maintain or enhance local accessibility', both of which simply double-count 'improved transport performance'. An argument was employed that an equal weighting should be given to the four criteria and the scores on each criterion summed to give a result:

> Overall, if key transport, prosperity, liveability and sustainability objectives are given equal weighting, Strategy Option 4 appears to represent a superior strategy to Strategy Options 1, 2 and 3. While its environmental (i.e. sustainability) performance is inferior to the other Strategy Options, adoption of Conclusion 19[5] and implementation of the management measures outlined[6] ... would enhance its performance. (DoI et al., 1998a: S-16)

If we accept that transport, prosperity, liveability and sustainability should be weighted equally then Option 4 does come up with the best overall result. If we take a broader view, this approach must be questioned. 'Trading-off' degradation of the environment for improvements in the economy will further reduce ecological sustainability. Even if we accept that an EES should include the standard sustainable development categories of economy, society and environment, there appears to be no methodological justification for 'transport' being a separate criterion in the evaluation process. Transport would usually be a sub-section of either prosperity (i.e. economy) or the liveability (i.e. society) category. Adding it as an extra category creates a bias.

The necessary relationship between ecological sustainability and economic growth has been wilfully misunderstood. This is evident in the discussion about the 1992 National ESD Strategy – the 'most substantive effort in Australia to integrate environmental, social and economic goals into a single concept'. The principles in the ESD strategy include 'to protect biological diversity and maintain essential ecological processes and life-support systems', but they also 'explicitly recognize the importance of material wealth creation' (DoI et al., 1998b: 1–2). Here there is a lack of clarity about the relative importance of environment and economy. 'Material wealth creation' (i.e. the economy) cannot reasonably be included in a definition of ESD unless it is very explicitly subordinate to the need for ecological sustainability; otherwise it is likely to continue contributing to serious ecological damage (pollution, resource depletion, biodiversity loss, and so on). Such an outcome is highly problematic, hiding the need for the fundamental economic

---

5 Conclusion 19 suggests an alternative, more expensive interchange in order to retain remnant vegetation (DoI, 1998: 11-63).

6 'Management measures outlined' simply include the usual measures to include the needs of cyclists, and pedestrians; to replant with indigenous vegetation; to control runoff; to construct noise mounds or barriers; and so on (summarized in Table 11.9, DoI, 1998: 11-44 – 11-49).

change required for ESD, that the economy must be restructured within the limits of ecological sustainability.

A Panel of Inquiry into the Scoresby EES appointed in September 1998 was chaired by Professor Roger Eade, who later argued that project-oriented Environmental Impact Assessment (EIA) procedures and methodologies were fundamentally inadequate to assess the degree of complexity found in the Scoresby proposal (Eade, 2000). According to Eade, project-based EIA 'forecloses on the proper consideration of alternatives to the project under consideration … an alternative put forward by the Public Transport Users Association which provided for a public transport solution to the transport deficiencies in the study area deserved full evaluation but in fact it was discarded early in the process' (Eade, 2000: 516). Eade suggests that a Strategic Environmental Assessment (SEA), which is applied both earlier in the process, and has a wider scope than project-based EIA, might help avoid this problem in future projects of such complexity (Eade, 2000: 516).

Sydney: Western Sydney Orbital

The Western Sydney Orbital is the western section of a ring road around Sydney (now called West Link). The main argument for the motorway is contained in the Environmental Impact Statement (EIS) (DOTRS, et al., 2000). The stated purpose of the EIS is, 'to assist the RTA [Roads and Traffic Authority] in determining whether or not the proposal should proceed'. Its 'objectives' include: 'discuss alternatives to the proposal and justify the selection of the preferred proposal' (DOTRS et al., 2000: 2.3). However, in marked contrast to the Melbourne case (which, however, did not include a public transport alternative); it appears that in Sydney no alternatives, even of road configurations, were considered even though the same consultant was employed in both studies.

The main considerations supporting the construction of the ring road were: ecologically sustainable development (ESD), connectivity by high speed road; freight transport efficiency, urban transformation, and, cultural icon (sic). The justification for the proposal is couched to a significant extent in the language of sustainability, and includes the mention of the 'precautionary principle', 'intergenerational equity', 'conservation of biological diversity and ecological integrity', and 'improved valuation, pricing and incentive mechanisms – namely, that environmental factors should be included in the valuation of assets and services' (ibid.: 39.3–39.4.). The EIS concludes its assessment with the claim that the proposal is in accordance with the principles of ESD (ibid.: 39.5). How this can be is discussed below.

Two storylines used are comparable to the Scoresby case, improved transport performance and increased economic prosperity. For the former storyline, 'connectivity' is the key term – and precedent: the need for the road having been identified in a series of past plans for the city:

> It would connect all of Sydney's existing motorways (the M5, M4 and M2) allowing high standard access to parts [sic] of Sydney (Roads and Traffic Authority 1998a and b)'. (ibid.: 8.1)

> The need for a circumferential highway through the western areas of Sydney was identified in the early 1970s along with the establishment of a road corridor to facilitate such a road. (ibid.: 24.1–24.2)

The latter storyline used the appeal of economic viability (along with the preservation of amenity on local roads) as a very strong storyline driving the argument for the road. Here it was noted that the primary purpose of the proposal is for: 'the safe and environmentally responsible movement of goods by separating this movement from surrounding residential development' (ibid.: 36.33). It was also argued that there would be cost savings from the new road:

> The level of road upgrading works required on the existing base road network would be greater if the proposed Western Sydney Orbital were not built. Construction of the proposal should therefore result in significant cost savings on works required for the base road network in Sydney. (ibid.: 7.5–7.6)

The economic tale is further reinforced by appeal to probable employment creation: 'Accessibility is one of the most important factors in industrial location … [the road] would also stimulate the generation of further investment and new industries in western Sydney' (ibid.: 4.13).

Under the 'urban transformation' story there is a sense that there is an unstoppable process of urban development in Western Sydney and that the road is merely a consequence of that unstoppable process:

> It should also be noted that the general character of the region is undergoing significant transformation owing to rapid urban development. The proposal would be only one element, albeit a significant one, in that transformation. (ibid.: 17.10)

Finally it was suggested that the road would be a 'cultural icon', 'The Western Sydney Orbital would be more than a motorway. It would be a major defining element along the western edge of much of the Sydney metropolitan area' (RTA, no date: 15). This claim was not present in the EIS itself but its presence in an RTA public document merits comment. Cultural and geographical identification with a motorway (the WSO) is clearly considered a positive feature by the RTA.

What we see in the EIS is, like the Scoresby case, a slippage of logic such that the idea of 'ecological integrity' has been merged into a very general concept of 'environment', so general, in fact, that it covers the entire surrounding social context. The final sentence of Chapter 36 (entitled 'Summary of Environmental Impacts') reads: 'Overall, the proposal is expected to largely result in positive cumulative environmental impacts and provide many benefits for the population of Western Sydney' (DOTRS et al., 2000: 36.39). But under the 'environmental impacts' listed in this summary we find: accessibility, traffic efficiency, safety, reduction in local traffic noise, economic growth, and employment generation. None of these (except perhaps noise in a minor way) are relevant to maintaining or improving ecological integrity, so they should have been considered as social rather than environmental impacts.

Returning to the more accurate description of ecologically sustainable development, anthropogenic climate change is obviously an example of both 'serious' and 'irreversible' environmental degradation, as well as being a genuine threat to 'ecological integrity'. But according to the EIS, construction of the road would increase carbon emissions only marginally. First it was argued that the road is for freight movement – but no suggestion was made that use by other traffic should be restricted – second that the induced demand generated by building the road (and therefore increased emissions) would be reduced by demand management measures and by an improved public transport strategy. It was noted that 'Managing the demand for private road transport on a metropolitan-wide scale is essentially a policy issue requiring a range of actions across government agencies'. All the road proposal can do is preserve 'opportunities for road based public transport' and support 'the initiatives of others (particularly the NSW Department of Transport in regulating road operators) to take advantage of those opportunities' (ibid.: 5.31). This looks like a classic institutional barrier, although it is somewhat disguised under the presumably well-meaning 'Integrated Action Package'. It is the RTA's role to build roads, and other actors' roles to mitigate the negative consequences of such road construction.

Past attempts in Sydney to restrict demand for car use and provide an integrated public transport service have had limited success (despite the use of the integrated transport storyline in contemporary planning documents discussed in Chapter 7). It seems overoptimistic to expect that such an outcome would necessarily follow the building of the ring road.

Perth: Roe 7

The Roe Highway forms the southern half of a ring road around Perth originally designated in the 1963 *Metropolitan Region Scheme*. The road was divided into eight construction stages but no financial commitment was made to complete the full ring road at that time. The first stage, 'Roe 1', opened in 1984. Each stage provided a connection to the greater strategic road network.

As in Sydney, the 'connectivity' storyline together with an argument for the need for the road based on past plans was evident. For 'Roe 7', considered in 2004, its designation back in 1963 was considered sufficient justification (Roe 7 Alliance, 2004). Partial funding for stages 4 to 7 was secured from the Commonwealth Government, providing \$76 million of the \$187 million cost (Government of Western Australia 2002, see also Chapter 4). At the opening of stage 4 the State

Minister noted that this funding would allow the government to 'to reduce the impact of freight traffic in Perth's southern suburbs' (Government of Western Australia, 2002). Roe 7 was evidently a foregone conclusion in 2002, before an assessment process had even commenced. This may explain why planning documents lack any robust analysis and justification for the road.

The Environmental Protection Authority (EPA) made the decision to assess Roe 7 using a 'Public Environmental Review' (PER). This was one of five levels of assessment – a PER is a less onerous assessment than an 'Environmental Review and Management Program' (EPA, 1986). The EPA also determined that a Public Inquiry was not needed. Under the PER process the proponent prepares an environmental scoping document followed by a report on impacts. Both outputs are agreed with the EPA. A period for public submissions on the report follows and again the proponent is charged with the responsibility of preparing a summary of 'pertinent issues and matters raised in submission' together with their response to them (EPA, 1986). The EPA assesses this report and advises the Minister.

In its PER report, the proponent, Roe7 Alliance[7], argued for the road. We see similarities with the Melbourne and Sydney cases in the focus on how Roe 7 would improve transport performance (particularly for freight) and economic prosperity. It was argued that Roe 7 would provide an efficient traffic connection and reduce the impact of heavy vehicles and traffic flows on other roads. It was noted that the south-west metropolitan areas was experiencing strong growth and, with the Perth CBD remaining as the economic hub, this would result in the higher demand for travel.

Again, as in the other two cases, environmental impacts were mostly[8] concerned with local social impacts rather than ecological integrity. Impacts discussed included noise and air quality on surrounding development. The Alliance simply asserted that, 'gaseous emissions do not adversely affect the environment or health, welfare and amenity of nearby land users by meeting the statutory requirements ...' (Roe7 Alliance 2004: 95). However the PER does not provide data on air quality management, noting instead:

> MRWA...negotiated a Memorandum of Understanding (MoU) with the DEP regarding air quality assessment requirements for road proposals' (ibid.: 97); and, the regional level component of this investigation has not yet been undertaken. The regional investigation will, however, be completed during the PER review period, with the results being reported to the EPA for consideration in conjunction with other submissions on the proposals. (ibid.: 97)

7 The Alliance comprised the State Agency – Main Roads Western Australia, Clough Engineering Limited, Henry Walker Eltin Contracting Pty Ltd and Maunsell Australia Pty Ltd.

8 Another impact was associated with the loss of a rare spider orchid, Caladenia huegelii.

This process of 'closed' assessment prevents public scrutiny of the environmental impact of the proposal. The MOU between MRWA and EPA also prevents public scrutiny, and renders the PER process impotent. The independent assessment of air quality impacts occurred during mid-2004, with reports made available publicly. There was, however, no opportunity for the public to comment on the reports.

The proponent's summary of 'pertinent issues' from submissions to the PER mainly focussed on how the proponent planned to manage the impacts from the construction and operation phases of Roe 7. This was in line with the scope of the PER which only addressed management and construction issues and not the wider justification for the road. Community opposition included submissions questioning the need for the road, including concerns with diminishing oil availability, promotion of only road-based transport rather than other options, and that traffic projections did not justify the need for a road for freight vehicles since they only make up a small proportion of traffic (4% of all vehicles). These concerns were not addressed and the road was constructed.

The overarching argument evident for Roe 7 lies in the 'rational planning' storyline: 'The construction of the Roe Highway Stage 7 will enable realisation of a ring road system that was first included in the 1955 Stephenson-Hepburn Plan for the Perth Metropolitan Region. The primary regional road reservation for the Roe Highway has been included in the MRS since its promulgation in 1963' (ibid.: 59). Yet the historic provision of a road reserve should not remain an adequate justification for the road's construction some forty years later. Significant environmental and social changes occurring since the 1950s surely warrants consideration against the policy environment of the time.

In all three cities we see the decision to proceed with continued construction of orbital motorways as part of a wider plan to connect the arterial road network based on a policy position justified in the late 1950s/early 1960s. This happened despite decision processes occurring during the turn of the century in three cities where we, earlier, noted some limited evidence that the policy environment was maturing as ideas of ecological sustainability were played out in key transport and land use policy documents. It appears, however, that the storyline of 'sustainability' had to be built into the justification of road building because it had entered the wider political discourse within the public sphere. Yet the reality of a truly sustainable transport system could not be grasped. Rather 'sustainability' was fitted ever more uncomfortably and illogically into a path dependent road building mentality supported by a nexus of economic and engineering storylines.

These cases provide some evidence of the power of the institution: seen in the way in which the roads agencies play centre stage while the public transport alternative rarely appears, and in the way decision processes themselves favour road building proponents. In all three case studies the Engineer's Tale and the Economist's Tale are strongly played – with justification based on transport efficiency, economic efficiency and the need to keep freight moving.

Conclusion

The critical observation here in this chapter is perhaps not so much how the policy discourse, and its associated tools and techniques, reinforces the implementation of a roads based transport system – one designed for high speed private car travel – but how the policy discourse plays out for public transport. Planning for public transport is the Cinderella of transport planning. It is not so much that Cinderella is absent, but much more that she is denied equality in the transport discourse. Public transport is consigned to the scullery on the basis that she does not provide a worthy alternative to the car. Modern 'man' desires a car and his wishes should be met. Public transport is a welfare option, and even if it were optimally provided it would never compete with the car in terms of speed or comfort (overlooking of course that public transport can and does compete in other parts of the world). The Cinderella metaphor concludes when we see towards the end of the century that Cinderella can indeed go to the ball – public transport is, belatedly, seen as a viable part of the transport system (at least in a published policy sense). Public transport can and does compete effectively with the car; it provides a transport choice for car owners. In Australian cities, now growing close to the largest in Europe, a public transport solution is the only option for the avoidance of terminal congestion. In the next chapter we turn to the discourse of public transport.

Chapter 9
The Discourse of Public Transport

As we found in analysing the roads discourse in the previous chapter, there has not been a single source of policy documents for transport in any of the three cities. Those documents selected to capture the transport policy over the 60 year period until the turn of last century were shown heavily to favour the provision of both transport infrastructure and urban development in support of travel by private car. For a long period public transport was seen only as a welfare option to be provided for those with no access to the car. In this chapter we search deeper into published policy in order to investigate the discourse of public transport, in this case focussing mainly on documents produced by the public transport agencies.

We apply the same research strategy for this investigation – examining the discourse through storyline analysis. We will see that the dominant public transport storyline from the 1960s has been 'containing the public transport deficit' – a drain on the public purse. This storyline continued right until the turn of the century in Melbourne and Sydney, but tailed off by the end of the 1970s in Perth. The storyline is tied in with the 'inevitability' story, that because the car is the modern method of travel it is inevitable that it will replace public transport and that little can be done by public agencies to counter this trend. Over the period there have been shifts in the dominant storyline (public transport deficit), at least in Perth and in Melbourne, towards a story that emphasizes world class public transport. Yet we shall see that in all three cities, as far as policy discourse in concerned, the feeling of inevitability of car-borne transport retains its hold, and public transport agencies still do not see their role as competing effectively with the car.

As before, we present our data in such a way as to do more than simply describe the presence, absence and frequency of the storylines, but rather, we seek to show how the storylines work to construct meaning and power relations. We used expert opinion to draw a purposeful sample of public transport policy documents that were produced in the last sixty years of the last century (Neuman 2003; Patton 2002) of influential government policies, plans and strategies. Added to this we sampled annual reports; this provided further insight into the policy position of public transport agencies where a clearly badged public transport policy was absent.

Melbourne

In total we drew on 52 policy documents, mainly the Annual Reports of the MMTB, VicRail, MTA, and the PTC. We cover the period from 1965 up to the eve of the new century in 1997.

The public transport 'deficit' was the major storyline of Melbourne public transport from 1965–1982. The deficit was the difference between total public expenditure and revenue from fares and other sources. This storyline was evident for all three public transport modes: trams, rail and bus. In 1965, the MMTB was recording an accumulated deficit which was attributed to rising wages, workers' compensation insurance, electricity and fuel costs (MMTB, 1965: 10). It was argued, however, that this 'deficit' was not in fact a genuine deficit; MMTB requested re-imbursement for the costs of concession fares – 'a cost to the Board … for what is in fact a social service benefit' – as well as 'making and maintenance of the roadways on its tram tracks', which saved the relevant municipalities large sums of money (ibid.: 11). MMTB had not historically received government subsidies: 'The Board is unique … in standing on its own feet financially … but this is no longer practicable under present conditions' (MMTB, 1966: 9). This line of argument was repeated each year through to 1970 (MMTB, 1966: 9; 1967: 8–9; 1968: 10; 1969: 7; 1970: 8). Finally in 1971 the state government began making such 'social service' payments, and has continued to do so, but by this stage inflation of wages was such that the Board continued in deficit despite this change. Defining payments to maintain public transport as 'social service payments', of course, reinforces the storyline that public transport is a form of welfare.

VicRail, like the MMTB, argued the need to be reimbursed for social services provided to the public, in particular: travel concessions for students and pensioners, and freight concessions for 'country industries'. Such concessions, it was argued, 'should not be submerged in railway accounts but should be separately financed … This would have resulted in a surplus … instead of a deficit' (VicRail, 1967: 9; VicRail, 1968: 11). What is interesting is that VicRail appreciated the 'lack of level playing field' early in the period. In comparing spending between roads and rail they drew on a quotation of the British Railways board of the previous year:

> The Minister is buying more than a railway passenger service for those who wish to use it. He is also buying benefits for a far greater number who are not using it, in the shape of less delay and congestion on the road and in parking spaces and in less deterioration in the local environment. It is not in fact the railways who are being subsidised through the grants but the railway passenger and the road user, especially the motorist. (VicRail, 1970: 2)

Aside from the 'deficit' storyline, the MMTB did engage with the 'congestion versus free movement' storyline. Traffic congestion was a major concern for MMTB throughout the 1960s and 70s, when the Board argued persistently for priority for public transport, 'in the public interest, it is indefensible that trams should be held up short of their stops by motor cars, many or most of them carrying only the driver, needlessly driving on the tram tracks' (MMTB, 1965: 13). Public transport, it was argued, was the only solution to traffic congestion:

> urban public transport is an excellent community investment. Overlooking for the moment the aged, the young, the infirm and the relatively impecunious who cannot drive themselves, but must still travel, if the remainder tried to use private motor cars, no city's street system could accommodate them; nor could any city centre provide space to park their cars and leave any city worth coming to. No one perhaps has a greater real interest in maintaining public transport than the private motorist: so long as others will continue to travel by train, tram or bus he is still able to drive his car, albeit in increasing. discomfort. (MMTB, 1967: 9)

By the late 1960s, the MMTB was partially blaming congestion in city streets, which frustrated movement of trams and buses, for the decline in tram and bus patronage (MMTB, 1968: 7), 'Disgruntled passengers take to their motor cars instead – and make the congestion worse still' (MMTB, 1969: 7).

Other storylines occasionally appeared. In the early years it was clear that the MMTB accepted that 'low density can only be served by roads'. The Board thought that plans to increase density in the inner areas would offset the decline in patronage (MMTB, 1965: 7) and referred to the growth of the outer suburbs as a reason for decline in patronage (MMTB, 1968: 7). The 'rational planning' storyline also appeared:

> Great cities the world over have recognized that even if, on a simple income versus cost basis, their public transport is unprofitable, it is indispensable for a go-ahead city and therefore must be supported and maintained, including supported financially if necessary. (MMTB, 1967: 8) (The example is given of the extension of the London Underground.)

> But planning for public transport was never to be at the expense of providing for the car, It is encouraging to find [that great cities overseas realize] that – without necessarily looking for any reduction in the present number of motor cars – fast, modern, attractive public transport is essential if the city is to continue to prosper (MMTB, 1969: 8)

This situation of ongoing accumulated deficit substantially limited any further investment in infrastructure provision, thereby reducing the quality of existing services. Feeder bus services were seen as a means of connecting residents in outer areas to the heavy rail network (MMTB, 1967: 7).

For railways there was promise of Federal assistance. This went some way towards a solution but was not enough. (VicRail, 1973: 2) VicRail continued to argue against the 'deficit' storyline, this time in relation to arguments that increased fares would reduce the amount of the subsidy. VicRail now drew attention to the need to compete with the car:

> It is now generally accepted in heavily motorised communities throughout the world that if passengers are to be induced to use public transport in preference to

> the private car, charges cannot be set at a level that will yield a cost recovery rate better than about 50 per cent; the average cost recovery rate recorded by 63 urban passenger undertakings which are members of the International Union of Public Transport was 55 per cent in 1976, after declining steadily for the previous eight years. The Victorian figure of 49 per cent is accordingly consistent with World experience. (VicRail, 1979: 5)

1983 was a critical juncture: the government amalgamated the MMTB and the urban arm of VicRail, creating the MTA (see Chapter 6) culminating in the production of *MetPlan* (MTA, 1988). The storylines changed significantly. While the 'deficit' storyline remained, other storylines became stronger. A new storyline appeared: 'the people's choice' – world-class public transport for Melbourne's future (MTA, 1985: 6, 7; 1987: 10, 19; 1988: 12–13). The public responded well to the integrated public transport system co-ordinated by the MTA, and from 1983–1987 patronage increased by 14% – after 'decades of falling patronage on public transport in Melbourne' (MTA, 1987: 19). This dramatic improvement' was attributed to 'the modernisation of public transport which has brought it up to a standard of which Victoria can be proud …' (ibid.: 19). The MTA continued to argue for investment, '… the services had to be made responsive to the needs of the communities in which they operated … To achieve this long range planning, investment is needed' (ibid.: 16).

The release of *MetPlan* in 1988 continued the 'world-class public transport' storyline, tying public transport in with land use solutions:

> A clear commitment to public transport has replaced an atmosphere of decline and decay in public transport. Now, we have the opportunity to plan for the future with confidence. MetPlan is the Government's plan for the next 15 years of public transport. It envisages substantial growth for the system … serve existing routes and cater for expanding areas in Melbourne … the most up-to-date trains, trams, buses and light rail vehicles in Australia …
>
> Commercial developments will be encouraged around stations to make them community hubs providing a range of service to Met customers in shopping developments … Switching from one form of transport to another at safe, clean modal interchanges will be easy. (MTA, 1988a, 1–2)

MetPlan set a target of 20% growth in patronage.[1] Moreover a, 'goal will be to provide and present a viable alternative to travel by car, removing the need for households to buy a second car' (ibid.: 8). Visionary it was, but the plan alerted the reader to the lack of funding to achieve the vision:

---

1 A target that was used again in 2002 in the *Melbourne 2030* plan.

> availability of funds for public transport in Victoria will not be increased and may be reduced in real terms over the next decade. Consequently *MetPlan* has been developed within the framework of no real increase in resources being available. (ibid.: 38)

This short period in which the discourse shifted immediately followed the election (in 1982) of a Labor government (under Premier John Cain) after a period of almost 30 years of conservative governments. The new government came to power with an agenda for restructuring the entire machinery of State government to put Ministers firmly in charge of their departments (Gleeson and Low, 2000; Laffin, 1987). The government created a Victorian Transport Directorate (VTD) 'to apply co-ordinated management to the total transport task' (Victoria, Minister's second reading speech, 1983: 4324). The MTA was one of four new authorities co-ordinated by the VTD. The intention was for the MTA 'to draw together Melbourne's public transport system into a single co-ordinated unit' (ibid.). The restructuring seemed to promise major rethinking of the urban transport task, and there followed some progress in increased investment in new (railway) rolling stock and the introduction of integrated ticketing across the system. But the government was also dedicated to fiscal responsibility and modern management which gave it a strong focus on financial control and deficit reduction. The Minister's speech introducing the Transport Bill was entirely concerned with administrative structure and contained nothing of substance about transport policy: for instance what an integrated transport system would be like or what needed to be done to achieve it.

The MTA's vision was short-lived and what followed saw further structural reform in the public transport agencies which distracted the focus away from competing with the car. In 1989 the MTA and V/Line were amalgamated into the Public Transport Corporation (PTC). The restructuring flagged in MetPlan (including redundancies for staff in stations, tram and bus conductors) had met with resistance,

The Public Transport Corporation's first year was something of a baptism of fire:

> While several major reforms were planned for the year, including dramatic changes in the ticketing system (MetTicket), these reforms were resisted by unions and a number of community groups and were not successfully implemented, resulting in severely disrupted services and the erosion of public support for the PTC. Financial targets were not met and gains in patronage earned over the previous five years were lost … The 1989/90 financial year was completed therefore with public support and patronage at their lowest level since the mid-1980s. (PTC, 1990: 6)

The 'deficit' storyline re-entered the foreground. Holmes (1987) comments on transport policy in Victoria, 'One major difficulty is that welfare states expect their public transport systems to supply social as well as commercial services, and the present crippling level of system debt reflects these expectations. A State

Opposition study released in June last year claimed that the railway system alone lost AU$1000 a minute, and indebtedness had increased by 197% since 1981–83'. In 1991 the PTC and private bus companies had 'agreed to work jointly to rationalise metropolitan services' to make AU$11 million worth of savings (PTC, 1991: 9). In 1993, under a new government with a neo-liberal agenda, the 'Public Transport Reform Package' was released stating that, 'The first essential step was to cut the huge transport funding deficit' (PTC, 1993: 4). More than 4,000 staff left the PTC in the 1992/93 financial year (PTC, 1993: 3), and over the following years the Annual Reports regularly report further staff reductions. In 1997 preparation for privatization was put into effect. Privatization of public transport operations was to be the solution to delivering a public transport system without deficit, 'Public transport in Victoria has reached a watershed … The shape of the PTC will change radically in the next 12 months. The result will be stand-alone, self-sufficient organisations, ready for privatisation'. (PTC, 1997: 5)

On the 'sustainability' storyline, the earliest reference to the environmental benefits of rail transport is a note about the potential of trams to 'provide low cost mobility to the citizens of this city without in any way increasing pollution hazards. This is a very important feature of the tramcar' (MMTB: 1972, 6). There was reference to the problem of air pollution at the time, which VicRail also highlighted with a poster. The caption on the poster read: 'Trainsportation (sic) is the silver lining to one pollution problem … fifty new stainless steel non-polluting electric trains are being built for Melbourne's metropolitan service' (VicRail, 1973: 3). In 1988 the MTA noted:

> if The Met doubled its fares, it is estimated that more than 100,000 extra cars would be on the roads in peak periods, most on roads leading into the city centre which could only be widened at great cost and with major damage to the urban environment. (MTA, 1988: 17)

But overall there was rare use of the 'sustainability' or 'environment' storyline by the public transport agencies.

Sydney

In total we drew on 27 policy documents – primarily the Annual Reports of The Urban Transit Authority, State Transit Authority and Public Transport Commission (see Appendix 2).

As in Melbourne, the 'public transport deficit' storyline was very influential in Sydney throughout the period. Each of the early public transport providers referred to large and growing deficits over the decades. The Commissioner heading the Department of Government Transport (CGT), which operated bus services in the 1960s and the early 1970s, blamed the downward trend in its financial results on rapidly declining bus patronage and the consequent reduction in passenger revenue. The Commissioner suggested that the patronage loss was due to the effect of expanding motor traffic, the development of shopping centres in suburban areas, and (curiously) 'increased growth of television' (CGT 1960: 2). The use of emotive language in discussing the deficit was not uncommon. In the '*Bus Task Force*' report of 1973 the deficits incurred by Government bus services were described as 'catastrophic' (Public Transport Commission 1973: 2). The Government rail services also experienced a hefty deficit but this was attributed to the Department's obligation to provide services which 'equated neither to traffic needs nor a realistic fare structure' (CDR 1972: 1). Here we see the 'public transport equals welfare' storyline. After the rail and bus agencies merged to form the PTC the 'deficit' storyline remained. The reports often referred to the need to restrain 'escalating over-expenditure', contend with 'crippling operating losses' and contain a 'rapidly rising deficit'. The deficit was attributed primarily to rising wage costs and a shortage of up to date equipment. The Commission claimed that, due to lack of capital funds, it was forced to use equipment and facilities which were 'worn out, obsolete or inadequate' (PTC 1973: 4). This confirmed that while Sydney was prepared to invest in a modern road system at that time, the same did not apply to the public transport system.

From 1981 onward rail and bus services were once again split up to be provided by two separate agencies – Urban Transit Authority – then State Transit Authority after 1989 (STA) and the State Rail Authority (SRA). A top priority for the STA was to achieve commercial operation and eliminate the need for deficit funding (STA 1995: 5). By the end of the decade (somewhat later than in Perth – below) the way in which costs were assessed changed, costs for providing 'non-commercial services', and pensioner and student travel, were met by payments from Government. The STA congratulated itself for its increasingly successful efforts to become 'less of a drain on taxpayers' money' (STA 1989: 5) as a result of this new arrangement. The SRA was subject to similar financial restructuring.

What is interesting is that the SRA did not buy into the 'deficit' discourse. They refuted the 'widespread community belief' that railway debt was a reflection of inefficiency and claimed that the deficit had grown as a result of past Government policy which required the railway to borrow money to fund capital works and provide services that could not be provided on a commercial basis. The Government accepted this and the shortfall between operating cost and fare-box revenue on 'non-commercial' SRA services was funded by way of Government grants (SRA 1990: 17).

Traffic congestion was a major issue for bus transport providers in Sydney and the discourse around the storyline 'chronic congestion versus free movement' was clear and often tied to reasons for the 'public transport deficit'. As early as the 1960s providers saw the main cause of traffic congestion to be, 'the continuing widespread use of the automobile for private transport' (CGT 1965: 5). Interestingly they noted that town planning ideas of concentrated development (which would later be argued by others as a way of supporting public transport patronage), such as the development of 'multi-storey buildings' in the City, would 'generate additional traffic both vehicular and pedestrian'. So bus services would remain unreliable until effective public transport priority measures were taken in the City (ibid.: 5). At least here they saw the potential for traffic management. They continued to express a concern about congestion in the 1970s: 'the volume of private cars and other traffic using many roads is reaching saturation point … resulting in disruption of bus services' (CGT 1972: 8). In a similar vein the PTC also argued for rail improvements on the grounds that this could relieve congested roadways by carrying peak hour commuters. For example, the Eastern Suburbs Railway would 'provide some relief to the city's road arteries by attracting commuters who may otherwise drive their vehicles' (PTC 1973: 20).

By the late 1980s the agencies argued that their 'principal objective' was to 'help to reduce traffic congestion' by attracting more people to public transport, but admitted that 'because of traffic congestion, Sydney's buses are the slowest in Australia' (STA 1989: 13), although they generally only argued for bus priority measures (UTA, 1981).

As the end of the century approached the 1998 integrated transport plan for Sydney, *Action for Transport*, continued the argument that public transport had a role to play in dealing with congestion, 'public transport must be used more to reduce traffic congestion and improve our air quality, but this cannot occur without substantial improvements to our public transport system' (DoT 1998: 3). Yet the plan argued for new high speed roads as the solution, despite the fact that this also facilitated private car travel. The plan claimed that the Eastern Distributor 'will not only get traffic off local streets, it will also result in significant improvements to bus services through a series of new priority bus lanes in currently congested streets' (DoT 1998: 11).

The first emergence of the sustainability storyline was in the 1970s. The 1972 report discussed the pollution potential of its vehicles, pointing out that, 'although the diesel engine is not considered a serious polluter either in terms of carbon monoxide or of hydrocarbon emission, under some conditions it can offend by emissions of dark smoke' CGT, 1972). The following year the PTC emphasized the need to make the public aware of the 'environmental pollution aspect and the accompanying benefits of the use of public transport' in its marketing efforts (PTC, 1973). In 1989 the STA framed its trials of alternative fuels as an environmental initiative saying that 'environmental issues are important in delivering safe, reliable and comfortable public transport. All advances in fuel technology are carefully examined and trials began during the year on the use of Compressed Natural Gas in two Mercedes-Benz buses' (STA, 1989).

What is clear is that the agencies mainly saw their contribution to ecological sustainability as a technical one where the vehicles they used were, 'technically efficient with a minimum use of non-renewable resources and the least possible environmental impact' (NSW DoP, 1993). Only rarely did they see a role for public transport in offering an alternative to the private car: 'one of the greatest contributions State Rail can make to the environment is to encourage more freight and passengers onto rail, removing trucks and buses from the highways' (SRA,

1990); and 'protecting the environment is high on State Transit's agenda. By encouraging people to leave their cars at home and travel by public transport our aim is to reduce the volume of city traffic and the level of vehicle emissions into the atmosphere' (STA, 1995). But, as seen in Chapter 8, the public transport discourse did not gain the traction needed to promote the development of a world class public transport system for Sydney. The final strategy evaluated at the end of this period lamented:

> Sydney's population has continued to grow but so has our car use. Public transport use has also continued to increase, but car use is growing faster, threatening air quality. If we continue at this pace our air quality will deteriorate, affecting our lives and those of our children and grandchildren. (DoT, 1998)

And finally it appeared that the 'deficit' storyline would no longer hold weight:

> The government considers that transport resources cannot be driven solely by economic and financial considerations of resource allocation and full cost recovery. The important issues of social equity and accessibility for all Sydney's residents as well as the need to protect our unique environment must be taken into account. (DoT, 1998)

This latest plan lists the completion of the rapid bus only transitways, Stage One of the rail construction program and the Sydney Orbital road – a plan full of mixed messages!

Perth

We drew a sample from 57 potential policy documents (Appendix 2) by consulting local experts on which documents were influential. Added to this, the sample included all Metropolitan Transport Trust (MTT) annual reports. The Western Australian Government Railways Commission (WAGR) reports were reviewed periodically; the reason being that, for the most part, the MTT set the policy direction for urban public transport while WAGR acted as a service provider under direction from the MTT, a sample of these was therefore sufficient. In total we draw on 40 policy documents.

As in Melbourne and Sydney, the existence of a public transport deficit storyline preoccupied Perth's public transport agencies for the whole period. Although many early MTT annual reports did not refer explicitly to the deficit, the Trust routinely reported significant operating losses and the constant discussion was of the need to 'obtain economical operation' (MTT, 1959: 3), 'improve cost recovery' (MTT, 1982: 17) and 'rationalise services' (MTT, 1982: 9; 1983: 9). The Trust drew attention to the role of increasing wage and fuel costs to explain its financial woes, but the documents left no doubt that the primary reason for deficit was the enormous fall in patronage brought about by competition with private cars (particularly at nights and on weekends) (MTT, 1963).

It was not only the MTT who drew attention to the public transport deficit storyline. The Director General of Transport (DGT), charged with providing policy for all modes of transport, drew considerable attention to the size and growth of the public transport deficit. In a 1976 review paper for the Minister for Transport, the DGT delivered ominous warnings that the deficit would increase to alarming levels 'without either regular and large increases in Public Transport fares or massive reductions in services' (Office of the Director General of Transport, 1976: 7). In the 1979 'Bus and Rail Policy' the Minister for Transport claimed that the need for Government to fund a 'large general deficit' for urban public transport resulted in less money 'available to the community for other worthwhile projects' (Minister for Transport 1979: 28). When denouncing the 'deficit', in none of the documents reviewed did either the Minister or the Director General make mention of the financial and environmental costs which public transport provision serves to abate.

The 'public transport deficit' storyline was often coupled with other storylines. The 'public transport equals welfare' storyline was evident, with the MTT seeing itself providing a welfare service – mobility for those who remained without access to private transport. This was also put forward as a reason for the deficit, with the reports frequently complaining about the need to provide 'uneconomic' or 'non-viable' services for those in need. The reports persistently described weekend, night and 'fringe' services as a 'costly liability' or 'burden'. 'Low urban density can only be served by roads' was also drawn on by way of explanation for the organization's incapacity to provide low cost services. This played into the discourse for a car-based solution.

Up until the 1980s the deficit storyline was mostly accepted rather uncritically. During the 1960s, however, there is some evidence that the public transport agency understood the need for an integrated transport solution –that in order for public transport to be effective it would be necessary to consider the place of private car travel. In 1964 a WAGR report criticized the complacency of motorists (the lack of gender neutral language is apparent in this report as in many other transport documents) in ignoring the unrecognized costs of driving:

> Seemingly the private car owner has not yet been able to appreciate that the convenience and the prestige which he believes the use of his own vehicle confers on him, is needlessly affecting his own pocket and the pockets of others in the community, and that as each year passes with its increasing number of vehicles coming into the city, this cost will grow in considerably greater proportion than the growth in vehicle numbers. (WAGR, 1964: 6)

In 1968 the MTT suggested the introduction of on-street and off-street parking restrictions in the city would be a 'step in the right direction' (MTT 1968: 4) and the following year suggested that 'such measures would not greatly inconvenience the community and would, in the long term, allow public transport to operate more efficiently and provide more attractive services' (MTT 1969: 5). The MTT also suggested that

> Town Planning Authorities should also give some thought to easing the present requirements for car parking space in high rise development areas. Some leading architects now consider that this type of development should be related to bus routes, with minimum on-site parking requirements. (MTT 1970: 3)

The stronger discourse from the public transport agency for an integrated solution was short-lived. The DGT retaliated against measures to restrict car use, stating that they would be 'counter-productive in terms of economic activity' (ODGT, 1976: 5). The MTT published an annual report at about the same time as the DGT review paper stating that while the Trust did 'not generally favour restrictions in respect of the motor car' it was nevertheless concerned about the increasing number of vehicles being accommodated in the city and in free car parking at shopping centres (MTT 1975: 4), although the report clarified that the Trust was 'not saying that provision of these shopper car parks is right or wrong' (MTT 1977: 3). After that the MTT fell conspicuously silent on the issue.

In the mid-1980s a visible shift was seen in the use of the 'deficit' storyline – the word began to be used in inverted commas as the MTT argued that this so called 'deficit' was not a genuine deficit at all but a reflection of the MTT's 'social welfare' functions (such as operating off-peak services to provide mobility for those without a car and providing affordable school bus services), 'The majority of public transport users are to some extent "captives": public transport is their only realistic means of getting about. If fares were increased these people would have little choice but to pay' (MTT 1985: 6).

By 1983 the Government accepted this view and made a decision to reimburse the Trust for expenditure incurred in carrying out these functions with 'Social Welfare Payments'. This was done in order to divorce the Trust's 'social service functions' from its 'commercial areas of operation' and give a more realistic impression of the MTT's financial performance.

The 'Economist's Tale', then, played a very strong role in the public transport discourse, not only in relation to the so called 'deficit', but also in the discourse around the 'travel as opportunity' storyline and the way this further helped to support car-based travel:

> the life style of the urban resident of today requires almost unlimited mobility for both work and recreation. No known or proposed public transport system or technology can provide the same degree of mobility as the private car. (Office of the Director General of Transport, 1976: 10)

The discourse was at times strongly emotive, serving to position any who would recommend limiting car access in order to promote public transport as seeking to take away freedom of movement and self-expression, 'if the policy of reducing the need for road construction by the provision of bus services for the journey to work is to be pursued with conviction, draconian measures to restrict the use of the private car, will be necessary' (Office of the Director General of Transport 1976: 9). Statements such as this probably curtailed any discussion of restricting car based travel.

The 'inevitability' storyline played a powerful role in the public transport discourse amongst several agencies, and in this way helped to underplay the need for a world class public transport system. Instead the policy strategists assisted in creating favourable conditions for the private car. The early MTT documents conveyed a clear sense that public transport was a superseded technology and cars were the 'way of the future'. It was argued that people had a strong preference for car transport and the lifestyle it enabled and, in planning for future transport, this situation needed to be accepted. This served to justify road building projects as evidenced by the DGT's claim that it was

> not possible to devise a public transport strategy which will reduce the amount of money that must be spent on roads in the Perth region', and it would be many years before there was 'any likelihood that journeys made by public transport [would] start to increase at a faster rate than journeys made by car. (DGT 1978: 5)

Likewise in the 1979 Bus and Rail Policy, 'it should be recognized that in Perth, where 90% of all households have at least one car and over 33% have at least two cars, it is very difficult to devise a public transport strategy which will reduce the dominance of the motor car' (Minister for Transport 1979: 22). So we see in a short space of time since the 1971 Perth and Regions Transport Strategy which proposed Perth's extensive high speed road network, that public transport policy saw only defeat. Even at the turn of the century, despite Perth succeeding in developing new railway lines and highly patronized orbital bus services, this defeatist discourse remained evident:

> In reality, inter-suburban services will generally not be competitive with the motor car. Typically, 'choice' travellers have no reason to stop using private transport for these trips. Until suburban development patterns change, this situation needs to be accepted. (Transport 1998: 7)

The responsibility, the public transport agency argued, lay with the town planners. In fact this statement is at odds with what has been achieved in Perth – a very low density city by European standards has succeeded in running a well-patronized public transport system.

Using the 'chronic congestion versus free movement' storyline, the public transport agencies' early discourse simply helped to support a transport system that would gave private cars dominance. The MTT sang the praises of new urban freeways on more than one occasion and actually encouraged building more of them, 'the introduction of more Freeways in the future would materially assist to provide fast, efficient services which would be an incentive for more use of public transport and a consequent lessening of the traffic problem' (MTT 1960: 6). The DGT used the congestion storyline to support arguments for more roads funding, stating that the 'growing inadequacy of some parts of the road network' resulted from 'a paucity of road funds' (DGT 1978: 12). The need to facilitate public transport was actually used as a justification for building new and bigger roads. It was argued that traffic congestion resulted in 'lower than optimum bus utilisation' with buses stuck in traffic delays. The solution the DGT argued was to proceed with 'planned road improvements' to avoid higher operating costs for buses (hence an increased public transport deficit) (ibid.: 8). Moreover, the DGT argued that congestion made it difficult for the MTT to adhere to timetables and this would lead to a 'reaction against public transport' (ibid.: 5).The irony is that a freeway based solution would make public transport even less competitive with the private car – as would later be recognized.

One of the most important storylines in the discourse was 'fixed rail obsolete'. This served not only to justify public transport agencies destruction of parts of their network that could compete with the car, but also helped to facilitate a roads-based bus solution. Like the 'keep freight moving' storyline in the previous chapter, these new roads were rarely built to provide separate, high speed rights of way for buses. Only later in the century did this occur, and when it did it was usually coupled with providing extra car traffic lanes too.[2] The 'fixed rail obsolete' storyline was, like other storylines, coupled with the 'public transport deficit' storyline. It was suggested that the major reason for the public transport deficit was the continued operation of suburban rail which was 'by its very nature' labour intensive and unjustifiably expensive (ODGT, 1976). The DG strongly advised against any further 'expenditure' on the urban rail system on the grounds that this would only serve to increase the existing deficit.

Both the 'deficit' and 'fixed rail obsolete' storylines formed the discourse for the closure of the Perth to Fremantle railway line in 1979. The solution to the deficit problem was to 'rationalize' services, and convert urban rail reserves into road ways with public transport provided by bus routes (DGT 1978). The DGT noted that the abandonment of services on the line would allow the railway reserve to be converted more cheaply and easily into a 'high standard arterial road' (DGT 1978: 17). In the policy document which announced the government's decision, the Minister for Transport described rail as an 'inflexible transport technology' (MfT 1979: 17) and argued that the railway could not cost effectively meet the changing transport needs in the corridor. Oppositional voices were positioned as irrational:

---

2 The most dramatic example of this in Perth was in the provision of busways within the centre of the Kwinana Freeway in 1998. These busways would provide commuter access from Perth's southern suburbs to the Perth Central Business District. To avoid a reduction on traffic capacity at the Narrows Bridge, the bridge was duplicated, thus providing an extra three traffic lanes at considerable expense (MRWA, 1998).

> Put simply, Railway Union members like railways, those that ride in trains like them, and there is an ill-defined but quite apparent public feeling which can be expressed thus:
>
> 'Every other city is installing underground electric railways, so why shouldn't Perth?'
>
> This assertion is wrong. Over the last two years a number of very large cities which have much higher densities than Perth have after extensive study, deliberately opted to provide additional transit capacity by means other than railways on the simple score that they are now fabulously expensive to construct and operate. (ODGT 1967: 24)

The Minister for Transport echoed these sentiments in a subsequent government policy saying, 'unwisely people often compare Perth with other capital cities pointing out that if other States can electrify their urban railway systems then why cannot Perth' (MfT 1979: 18). Ironically, the Government's resolution not to continue with rail services was, purportedly, based partly on the mimicry of decisions made in other cities. And while railways might be 'fabulously expensive to construct and operate', the statement is value-laden and no mention is made of the costs of alternative transport solutions.

Nevertheless, despite a history of defeatism and opposition to expansion, or even in some cases retention, of railways, by the turn of the century Perth had achieved an impressive change in public transport policy when compared to Melbourne or Sydney. The policy discourse clearly changed. 1983 was a critical year. Burke's Labor Government replaced the Liberal government. Labor had pledged, in response to strong community opposition, to reopen the Fremantle railway line which had been closed in 1979. The pressure group, Friends of the Railway (formed by Peter Newman, then a lecturer at Murdoch University and local government councillor for the City of Fremantle), had run a strong campaign including mass rallies and a petition of 100,000 people (Newman, undated) arguing a case for re-opening. In 1983 the railway line was reopened and the urban arterial road was not constructed. At the same time it was agreed to electrify the whole railway system. This was a catalyst for the re-opening of the Fremantle line. In 1991, all of Perth's urban rail lines were electrified. The 1991 glossy cover of the Transperth annual report celebrated rail technology with 26 photographs of the new electric trains (Transperth, 1991). This critical juncture was politically driven.

The political support for railways continued during a period of Liberal government (1993–2001). There remained no real change in the attitude of the bureaucracy until several years later. For example, busways were still seen as technically superior to railways in terms of patronage and this continued during the deliberations over the type of public transport for the then new northern suburbs corridor in the late 1980s. Slowly the published policy discourse changed. The 1995 *Metropolitan Transport Strategy* marked a point where the bureaucracy embraced the notion of 'land use transport integration' and 'balanced transport'. While the strategy mentioned the relative cost of Perth's rail system, this was immediately countered with an acknowledgment of the quality of service and comfort afforded to rail passengers, Perth's urban rail was, 'the most advanced and integrated system as well as the fastest and most comfortable urban passenger train service in the country' (Transperth & DoT 1993: 10). A set of 'forces' had come together – the community backlash against the closing of the railway; Newman's role inside government when he was seconded to the Premier's office in the early 1980s; and the progressive institutional re-structuring brought about by Stuart Hicks (Chapter 6) with the objective of integrating transport. Another point which marked a key difference to the critical juncture in Melbourne was the strong advice by a public transport bureaucrat that public transport privatisation must not follow the approach seen in the UK and Melbourne. Instead government must retain control of service delivery.

The 'fixed rail obsolete' storyline almost disappeared from the discourse, although in the 1998 '10 Year Plan' it was noted that

> the electric train network has a very positive image, but the remainder of the system, in particular the bus services, is typically perceived as second rate and hard to use' (Transport 1998: 23). While busways do not have the same initial community appeal as trains, they have a number of advantages in a low density environment. Apart from the much lower infrastructure and rolling stock costs, busway systems do not have to rely on a separate feeder network, and they can be implemented gradually, section by section, where necessary. Busways also provide safe passage for emergency vehicles. (ibid.: 20)

The 'sustainability' storyline rarely featured in the public transport discourse, and where it did it did not advantage public transport. The extent of the MTT's discussion of environmental issues in the 1960s and 1970s was limited. They reported on trials of alternative fuels for many years, but this appeared economically motivated rather than by concern for the environment. One reference was made to the environmental effects of over reliance on car transport – the MTT contended that if maximum use were made of public transport then not only would this reduce traffic congestion but also 'the amount of pollution in the air' (MTT 1970: 4). Similarly, while the DGT acknowledged that 'cars do contribute to air pollution' it claimed that emission control measures would ensure that the 'absolute level of car caused pollution is not likely to rise as the numbers of cars increase' (ODGT 1976: 11). The Minister claimed that any excess pollution expected in the transition to bus-based public transport after the closure of urban rail would be alleviated by emission control equipment (MfT 1979: 27). In the early 1980s the MTT referred to the 'subsidiary purposes' of public transport by acknowledging the role of the system in enabling the community to reduce the undesirable side-effects – noise and air pollution, accidents, energy usage, congestion and non-optimal land use (such as car parks) – of the automobile (MTT 1982: 31). Throughout the 1980s the MTT argued in its reports that reducing these 'undesirable side-effects' of car transport was an important justification for the Government's public transport subsidy because a well-supported public transport system would mean '… lower pollution, petroleum energy savings, fewer car parks, bridges and freeways, lowering of automobile accident rates and human suffering costs' (MTT 1985: 13).

The sustainability storyline also came into play when Transperth argued for public transport priority measures. The provision of a freeway bus lane would result in 'a modal shift from car to bus, with its consequent environmental, fuel resource, accident reduction, and other community benefits could be expected' (Transperth 1986: 18). Several years later, in an optimistic tone, Transperth claimed that 'as community awareness of the environment increases the public will also become more aware of the positive contribution public transport can make to our city's wellbeing. There is then likely to be a conscious shift away from car usage in the years ahead' (Transperth 1991: 3). In the same year (1991) the report noted Transperth's support of 'World Environment Day' by providing bus-back advertising, releasing a pamphlet about the environmental benefits of public transport and giving out prizes to public transport users on the day (Transperth 1991: 28).

Case Study – A Major Public Transport Project

In our investigation on the roads discourse (Chapter 8) we examined three major road infrastructure projects to assess the extent to which storylines in published policy were applied in decisions about road projects. We apply the same approach in this chapter. What is a telling difference, though, is that there were no major public transport infrastructure projects occurring in either Melbourne or Sydney during our evaluation period. Only Perth offered this possibility.

Perth: South West Metropolitan Railway

In its 1989 State Platform the ALP undertook to examine the feasibility of a number of possible suburban rail extensions, one of which was an extension to Rockingham and Mandurah. In 1992, after a series of consultant's studies (Nairn, 1990; Bettison, 1992a, b, c; Travers Morgan et al., 1992) the Premier gave a commitment to this railway, indicating that a range of alternatives was being considered. In the mid-1990s a change of Government, while continuing support for the railway, changed its route to run from the south, diverting at Jandakot to connect with the Armadale line at Kenwick en route to Perth. This amended route was included in the Metropolitan Region Scheme in 1994 and funds allocated in 1995. A 'master plan' for the South West Metropolitan Railway was prepared in 1998 and endorsed by Government in 1999 (SWMTPSC, 1999). In 2001 a new Government was elected and decided to build a more 'direct' route along the Kwinana Freeway median. After heated public debate on the best way to bring the railway into the Perth CBD, the Minister for Planning and Infrastructure appointed the Perth City Rail Advisory Committee (PCRAC) later that same year to provide advice on the best route. In 2002 PCRAC recommended the 'central route', whereby the railway left the freeway after the Narrows bridge and travelled underground through the Perth CBD with stops at the Esplanade, William Street and Perth City Station (PCRAC, 2002). A supplementary master plan for the rail line was produced by the Perth Urban Rail Development Project in line with this recommendation. The railway eventually opened in December 2007.

The discourse surrounding the railway is found in a mix of consultant's reports, political party documents, reports from committees appointed by government, and the public transport agency master plans. There was no one dominant storyline. Instead we see a mix of 'balanced transport', perhaps driven by 'chronic congestion' concerns, together with a debate about city form.

The 1989 Labor policy platform provides the earliest mention of the railway using the 'balanced transport' storyline – but in this case arguing for redressing the balance with greater priority to public transport, 'previous transport planning has over emphasised private motor vehicles without regard for the social and environmental costs' (ALP, 1989: 246). To redress the balance, Labor promised to 'implement major strategies to reduce the serious adverse impacts of our dependence on private motor-vehicle usage' (ibid.: 247). These thoughts were later echoed in the consultant's study, which suggested that the 'desirable pattern' would be that the new infrastructure not only caters for the anticipated travel demand but also that it 'generate new trips and divert people from private transport' (R.J. Nairn & Partners 1990: 14). An economic argument was also used where it was suggested that better use of public transport would also, 'reduce pressure on valuable inner city land by the avoidance of road building' (ibid.: 247).

'Predict and provide' tended to be a storyline used more by the consultants (who also sat on the steering committee), thus the feasibility study was said to be, 'prompted by envisaged development in the South West corridor' (SCSWCTS 1990: 1). The feasibility study was largely based on a travel demand and cost/benefit study that said that the investigations into a possible rail extension had, 'come about due to rapid population growth in the South West corridor and the perceived need to increase employment in the area and also the need to improve access between the area and the Perth Metropolitan Region' (R.J. Nairn & Partners 1990: 1). Unlike the roads rationale, which relied on forecasting demand based on past use, it was clear that to make the case induced demand would be an important argument, 'benefits are not simply enjoyed by existing travellers who now have a more convenient, or quicker, journey. Some people will travel who did not do so before because it was not previously worth it. These new travellers will enjoy a range of benefits' (SCSWCTS 1990: 18).

The 'chronic congestion' storyline also appeared. The 1990 feasibility study mentioned preventing traffic congestion as 'an issue of considerable significance' and the consultants noted that, 'roadworks currently planned by the Main Roads Department are not sufficient to prevent a considerable increase in road congestion or deterioration in the road operating environment' (R.J. Nairn & Partners 1990:

29–30), therefore they intended to assess 'likely levels of demand and congestion on major [road] links as modified by the impact of the railway service' as part of their cost/benefit analysis.

Another important debate that took place in the early 1990s was about which kind of transport technology to use for the rapid transit way – bus, light rail or heavy rail. One of the key issues enmeshed in this discussion was about the vision of urban form for the development of the south west corridor: should it contain self-sufficient regional centres, or rather dormitory suburbs where people lived but commuted to Perth to work? This discussion on integrating transport with urban form was not present in the road case studies. The committee concluded from their investigations that heavy rail was the 'least favoured option' because 'the restrictive routing requirements of the heavy rail technology' meant it could not meet 'some of the key system and corridor development criteria'. Here they referred to the problems with heavy rail gaining good penetration into regional centres such as Kwinana and Fremantle which would limit opportunities for intensification of activities around the station. The committee cautioned that high speed, heavy rail coupled with park and ride facilities might encourage 'public transport "commuter belt" development and an emphasis on Perth'. They argued that light rail would encourage greater movement within and between centres, rather than through them to the Perth (SWCRTST, 1990: 61). They recommended a light rail system (ibid.: i).

Others advocated the heavy rail option. During a SWAT technical seminar the Commissioner Gill, argued that Westrail's modelling had shown higher passenger volumes for heavy rail than light rail, 'we have carried out some modelling of our own which suggests that Metro Rail would attract many more passengers than Access Rail. That is, the results disagree with the figures produced by the team led by Transperth' (Gill 1992: 2 – part of the SWAT Technical Information Pack, 1992). Journey times to Perth were seen as the most important determinant of patronage levels. The debate over alternative railway routes also focussed on competing with the car and travel time became the overarching factor in the final decision:

> What is the most important planning question concerning public transport? As the Minister for Planning and Infrastructure has said, the most important planning question is to get people out of cars and into trains. It is the most important question in Perth today. How does one answer that question? It is answered by making the rail system attractive by making it fast, safe and clean. People will not go from cars to trains unless the rail journey is quick… The Government has come up with a system that, from a planning point of view, will get people out of cars and into trains. (WA Parliamentary Debates, 2001: 1695–1706)

> The Government has always said that the central reason we needed to change the rail alignment was to provide a system that would be competitive with motor vehicle travel. That has been at the very heart of the courageous and enlightened decision the Government made to build a rail link directly down the centre of

> the freeway. Main Roads has recently done a detailed study of the car trip times between Mandurah, Rockingham and Perth, given the freeway extensions. The information is interesting. Once again it confirms the profound wisdom of our decision. At peak times it takes 68 minutes to travel from Mandurah to Perth on the freeway by car. On our fast, direct train it will take 48 minutes, a saving of 20 minutes. However, it would take 60 minutes to wind one's way via the Kenwick deviation, which would provide very little saving. (WA Parliamentary Debates, 2002: 1182d)

This is an interesting perspective, which mirrors the situation caused by high speed motorways. So even where major public transport infrastructure is implemented, the prime consideration is how to compete with the car which sees the implemented solution reinforce a pattern of urban development that the car has created.

In relation to the 'sustainability' storyline, there is a very limited mention of this in the published policy discourse. Instead the discourse focuses on using storylines opposed to those used by road infrastructure planners. Rare mentions include that, 'Ecological and social impacts of the system should be minimal' (SWCRTSC, 1990: i). And fifteen years later:

> My [MacTiernan] serious message today is that the phenomenon of fuel price increases is here to stay. There is no easy solution, no quick fix and no quick subsidy that will resolve this fundamental issue. We all have to get on board with the idea that we must build resilience into our community and make people less car dependent, and that we must be prepared to invest in public transport into the future. (WA Parliamentary Debates, 2005, 5603b)

Conclusion

The discourse of public transport agencies itself did, for a large part of the sixty year study period, play directly into the hands of roads based transport policy solutions. Public transport was seen only as a welfare solution, provided for those without access to the car. Providing for public transport was seen as financially costly. Eventually, in each city, it was realized that it was government's role in fact to provide such a service and social welfare payments were made to the agencies, but this was never enough to put public transport on a level playing field with publicly-funded private transport.

The obsession was with the need for unlimited mobility, as part of the lifestyle and opportunity discourse, and even public transport agencies argued that this could only be provided by the car. Rather than provide an alternative discourse to the Town Planner's Tale of low density development being unsuited to public transport, the agencies succumbed to this story and its inevitable solution of car-based transport systems whose operational costs are paid by individuals rather than by governments. Even where there was the possibility of supporting a public transport system that could compete with car speeds (rail), the agencies often joined in the discourse in support of road building as a means to provide a bus based system.

Overall it appears that public transport agencies have not had strong storylines of their own and where the strong storyline of sustainability could have been used, it has been used only rarely. Importantly, while public transport agencies have attempted to respond to some of the strongest road storylines: 'infrastructure investment leads to economic growth' and 'predict and provide', they have largely failed. Instead they have allowed themselves to be buried in the discourse around the 'public transport deficit' for many years.

Chapter 10
Stakeholder Groups

In this chapter we turn to examine the way in which different stakeholder groups have engaged in the transport policy discourse. We compare the pro-roads groups with the pro-public transport/anti roads groups in each city. Again, we use the storyline as an analytical tool to identify the way in which these groups articulate their message.

Stakeholders: Pro-Roads

The pro-roads groups command a particular power in the transport policy discourse by virtue of their longevity, their membership numbers and their income. Over time they have become embedded within the institutional structure for transport planning. They form part of the 'discourse network' of roads. As Low (2005: 53) argues, 'Once a discourse network is well established, there is no need for active coalitions or conspiracies. The web of storylines is enough to ensure that influential persons think and speak in harmony when it comes to discussion of what is, in reality, an extremely controversial policy'. Their consistent message has been one of the shortage and poor repair of roads. They rely on road safety as an emotive argument for investment in roads (though almost never for restraining car use), as well as the need to keep traffic moving for the economy's sake. These storylines function as discursive barriers to policies promoting sustainable transport in Australia. It is only at the turn of the century that we see arguments for public transport investment, this, in the face of oil depletion where the technical fix of alternative fuels may not come soon enough.

In this section we consider the discourse of two car clubs – the Royal Automobile Club of Victoria (RACV) and the Royal Automobile Club of Western Australia (RACWA). We compare their discourse with that of the National Roads and Motorists' Association (NRMA) in its 1990s 'Clean Air 2000' campaign which we selected because it represented a proactive approach which specifically addressed some of the institutional barriers preventing sustainable transport.

Melbourne: Royal Automobile Club of Victoria (RACV)

The RACV was established in 1903 to 'serve the Victorian community, promote and advocate for motor cars'.[1] The RACV now represents over 2 million members

1 http://www.racv.com.au/wps/wcm/connect/Internet/Primary/about+RACV/RACV+history accessed 25 October 2011.

and an RACV member can be found in 'two out of every three households in Victoria' (RACV, 2001). A content analysis of *Royal Auto*, the house journal, was made, sampling 10 of the years between 1980 and 2001. The sample of years selected coincided with the release of major State policy documents when commentary relevant to transport policy was likely to be more evident. Four core storylines were present throughout the period: road funding, 'the people's choice', fuel tax, and later in the period – public transport.

The need for increased road funding was a major storyline throughout the period. The most powerful variant of this storyline links road construction with the health of the economy. For example, on the front cover graphic of the June 1987 edition of a bridge finishing in mid-air and a truck falling off it, the caption reads, 'When road funding's cut, the economy goes the same way' (RACV, 1987a: 5). Accompanying this is the argument that roads are essential to the economy because

> they carry most of our workforce to its place of employment as well as 80 per cent of our freight and 83 per cent of tourist traffic; road building is labor intensive; new and improved roads would save commercial and private vehicle travel time, and reduce vehicle operating costs; increased safety on new roads would reduce expenditure on road accidents; and, reducing transport costs will help increase the competitiveness of Victorian industry. (ibid.: 5)

The storyline – that cars are 'the people's choice' and that they are fundamental to the modern lifestyle of individual freedom – was prominent. Again, this storyline was linked with the economy, 'Australia is one of the most highly-motorised nations in the world … Our standard of living and quality of life are interwoven with our ownership and extensive use of the automobile (RACV, 1980a: 3). Indeed the car was described as 'essential' because public transport cannot provide the service – a direct attack on public transport:

> the motor car is not a luxury in Australia, it is an essential part of the lifestyle of our population where much of the population is scattered in remote areas and where in the metropolitan areas public transport is not at this time a viable alternative … Public transport is suitable for scheduled limited periods when the numbers of travellers have a common starting point, destination or route. This leaves open significant areas of time and location which must be served by the private car. (RACV, 1980b: 4)

The RACV ties the 'people's choice' in with the Town Planner's Tale, 'Motoring: a right, not a privilege … Motoring is for many an absolute necessity, particularly as the urban sprawl places demands for commuters and road users well ahead of infrastructure's capacity to cope' (Johnson, 1999: 10). While it may have been the case that public transport did not serve some work journeys, the biased conclusion is that further road construction should be encouraged rather than addressing the root of the problem – improving the public transport service.

The Club's resistance to fuel tax featured strongly throughout the period 1980 to 1995 (*Royal Auto* cover illustration, April 1980; RACV, 1980: 2; *Royal Auto* cover illustration, March 1981; RACV, 1981: 2; RACV, 1981a: 2; RACV, 1987: 4; RACV, 1995: 3). This was despite the Club's awareness of the oil depletion issue and later the environmental concerns. RACV's earlier solutions were seen in the 'technical fix' storyline – developing alternative fuels, 'Since man will not give up his freedom or mobility at any price, the best solution seems to lie in developing fuels which can supplement or replace petrol for use in the internal combustion engine' (RACV, 1980c: 9). Any recognition by RACV of the negative effects of private cars led to it to conclude that cars simply needed to be safer and less environmentally harmful, rather than have their use curtailed. By the late 1990s, RACV recognized the environmental dilemma of car use, but considered it simply a case of good car maintenance – here reporting on a survey of members:

> Motorists say they want to make what goes into a car environmentally friendly and what comes out of it less polluting. *But they don't want to change their driving patterns* and want to continue using their cars at times when they see fit. There was spontaneous mention that a well-maintained car will help reduce its environmental impact. (Cumming, 1998: 7; emphasis in original)

This is a good example of what Dudley and Richardson (2000) call filling the 'empty chair' of public opinion.

By the turn of the century RACV turned to address the public transport question, not just because out of environmental concern but also as part of the 'chronic congestion' storyline, 'More than 90% of respondents indicated RACV should lobby government for a better public transport system in order to reduce congestion and improve the environment. Similarly, strong support was given for ensuring that motorists' needs are met' (Daly and Hannan, 2001: 25). It was clear, however, that they 'support public transport initiatives, but not at the expense of needed road improvements' (Johnson, 1999a: 15). Herein lies the problem, government funding realities mean that there is only a limited amount of money to be spent on all forms of transport. Roads and public transport have always competed for available resources. The RACV's promotion of the idea that 'we can have both roads and public transport' was a clever political move when roads have always taken the largest share of government investment funds available for transport. The RACV made clear that it did not want to see motorists being at all disadvantaged by any transition to forms of transport other than the private car, 'The RACV encourages [public transport, walking and cycling] initiatives but does not want to see a negative attitude towards cars and private users' (Sanderson, 1996: 13). The RACV concludes that

> we need to be realistic when we think about alternatives to the car and that – with due restraint – as a community we must take maximum advantage of the benefits the car offers. Some of the restraint will come automatically with more fuel-

> efficient, safer vehicles, while some may need to be imposed, such as charges for using congested or environmentally sensitive roads. (Lay, 2001: 4)

While RACV, latterly, advocated for a better public transport system, they argued that it must behave as a car – and here we see that it must be road-based, rail based solutions were not entertained:

> the challenges will be best met by road-based public transport – fuel-efficient small cars, powered cycles, cheap taxis and flexible bus services – coupled with the use of telecommunications technologies to ensure the systems operate efficiently and effectively ... There is also much to be gained by making it easier to for people to meet their transport needs by walking and cycling. As with public transport, the challenge here is to meet the levels of safety, security and comfort provided by our private cars. (Lay 2001: 4)

Perth: Royal Automobile Club of Western Australia (RACWA)

RACWA was founded in 1905, two years after its equivalent in Victoria. The driving force behind the formation of the Club was the 'shared enjoyment of the freedom and convenience of the motor car' and one of its key roles was 'advocating and protesting the rights of motorists' (RACWA 2000a: 5). The Club played an important role in shaping perceptions, for example, in discussing a RACWA campaign for better roads an objective was to increase their members 'perception that road funding deserves a higher priority and that it therefore becomes an election issue' (RACWA 1990b: 2). We draw on a content analysis of 268 issues of the *Road Patrol*, the official publication of the RACWA, for the period 1960–2005. Over these four decades lobbying on motoring issues was very similar to that of the RACV, and included roads funding, fuel tax reform, petrol pricing and road safety. Three core foci can be seen in the RACWA's discourse: the need for road building and improvement; justice for road users; and, later, sustainability.

During the 1960s RACWA looked to the 'chaotic' traffic conditions in Europe and the United Kingdom' for evidence of why it essential to plan for the motor car in designing Australian cities. In discussing the UK Buchanan report, it was claimed that Britain's urban environment was 'out of date' and that steps needed to be taken urgently so that the motor vehicle would not 'defeat its own utility' (RACWA 1964c: 7). A subsequent review of the Buchanan report (see Chapter 1) by the Club claimed that, although the car is an entirely beneficial invention, 'almost at a stroke, the motor vehicle has put our urban arrangement of buildings out of date. It demands quite different arrangements of buildings and access ways ... the motor vehicle really demands a new type of town altogether' (RACWA 1964b: 26). Another article quoted the view of the Chairman of the UK Royal Automobile Club that what was needed was a 'positive dynamic approach with more and better highways and adequate off-street parking and garage space at a reasonable price ... [and] a comprehensive parking plan and road construction scheme' (RACWA 1964a: 21). RACWA argued that because Australian cities were then in a relatively early stage of development they were well placed to create car-oriented cities. This approach was perceived to be far preferable to placing restrictions upon vehicles and encouraging people to use public transport:

> There has been speculation in the Press on the situation created by increasing traffic densities in the city streets. A suggested solution is to price the motorist out of using his own vehicle and thus divert him to the use of public transport. This problem is world-wide and in some of the older and larger cities, whose basic design permits of no alternative, the predominant use of public transport has long become accepted. Here in Perth the situation is different. Fortunately, there is scope for providing off-street parking for vehicles on the periphery of the city and, indeed, within the city core. (RACWA 1966: 4)

Ironically, in the following decade the RACWA argued that the layout of Australian cities had made it impossible for travellers to use public transport, 'the decentralised spread of Australian urban areas is forcing people to use their own vehicles to commute to and from work'(RACWA 1971: 1), 'the spreading of residential and industrial areas away from the city centres' resulted in complex and dispersed patterns of movement throughout the metropolitan area and this made public transport 'inappropriate for most urban dwellers' (RACWA 1979: 2). So here we have a circular argument: 1) we should design the city for the car; 2) the city's design (for the car) makes public transport inappropriate; so, 3) we should design the city for the car.

The RACWA consistently complained about what it saw as a shortage of roads funds in Australia, 'because of inadequate investment in roads and associated works, urgently needed projects throughout Australia are being deferred or left incompleted (sic) to degenerate into unsafe, pot-holed and fuel wasting tracks' (RACWA 1980: 1). An issue of particular concern to the club was the belief that Western Australia was being allocated an inequitable and insufficient share of Commonwealth Government roads grants:

> the shortfall in road funding cannot be justified. Money spent on roads brings measurable benefits in terms of economic expansion, increased employment, reduced road trauma, lower cost of living, better access to essential services, increased tourism, lower congestion and ultimately, a better environment. (RACWA 1995a: 18)

This quote captures a number of different storylines of which the Economist's Tale (infrastructure investment causes economic growth; travel as opportunity) and the Engineer's Tale (chronic congestion versus free movement) are both apparent.

Turning to the Economist's Tale first, the economic importance of investing in roads is argued throughout the whole period. In the 1960s the RACWA argued that poor roads adversely affected national productivity and that an expanded roads programme would bring a range of economic benefits to Australia. These benefits included: providing employment; reducing the cost of goods; strengthening our position in world markets; stimulating industrial growth, assisting in the expansion of export trade; reviving the immigration programme and attracting overseas investors (RACWA 1962a: 21). Two decades later the discourse continued, arguing that these benefits would far outweigh the costs of building roads, 'the economic benefits which result can be many times the cost of the road works' (RACWA 1982: 2). By 1995, 'The spending on roads is easy to justify. One Australian study indicates that each additional $100 million spent on highways will produce an average of about 1,700 new construction jobs and a further 2300 new jobs in other economic areas' (RACWA 1995a: 26). By the 2000s the discourse argued that investment in the roads must be maintained, 'we have billions of dollars invested in our road network; the community and our economy can't afford to let it deteriorate' (RACWA 2003: 31).

Road safety featured prominently in the RACWA campaigns aimed at increasing funding for roads. The organization believed that the Federal Government had overlooked 'the value of high-standard routes in solving traffic congestion and road accident problems' (RAC 1968a: 3). On a rare occasion one article recognised that road accidents were an inevitable consequence of transportation and mobility, but it nonetheless argued that increased funding for roads would reduce the extent of the problem, reinforced with a 'justice for road users' storyline based on the taxation of individuals:

> A certain level of morbidity and mortality is the price that society pays for mobility, but the dimensions of the problem can be greatly reduced if only the political will is there. All it takes is money, and lots of it! … The Federal Government extracts large sums of money from motorists via fuel taxes. Only a small percentage is returned to road users for funding of road construction, road maintenance or accident black-spot remedial measures. In this particular context, more money means better and safer roads for the benefit of our entire society. (RACWA 1988: 36)

In 1990 the RACWA launched a campaign for road improvements, aiming to convince both Federal Government and Opposition that Australia's 'current road system is unsafe and inadequate' (RACWA 1990b: 2). Several issues ran an advertisement for the RACWA supported 'Fix Australia, Fix the Roads' campaign which graphically depicted a car crash involving three vehicles and carried the headline 'stop the carnage – fix the roads' (RACWA 1994a: 61; RACWA 1994b: 65). The prevalence of this storyline did not diminish over time. Restated in 2004 is the belief that high standard roads 'are a key weapon in reducing death and injury on our roads' (RACWA 2002b: 6–7).

The Economist's Tale was also seen in the storyline 'travel as opportunity'. In an article titled 'Modern freeways a blessing not a curse', RACWA saw the benefits of urban freeways as providing, 'Increased mobility, reduced travel time and less cost for all road transport freight and commercial vehicles, buses and private cars.' (RACWA 1973b: 2). This was quite a contrast to the 'deficit' story played out in relation to public transport (Chapter 9). RACWA extolled the importance of the car for individual freedom, 'most vital concern is for the maintenance of private motoring and the continued ability of people to move individually and in freedom' (RACWA 1978: 2). RACWA argued that criticism of the car was unwarranted because, 'Critics have ignored or overlooked the essential role of the car in the movement of people for business, social and recreational purposes. They have, too, disregarded the obvious wish of most Australians to retain the right to own and use their personal transport mode' (RACWA 1978: 20).

RACWA engaged with the sustainability discourse around two main areas of concern: the impact of the car on the environment and oil depletion. Environmental pollution was raised several times during the 1960s and 1970s and some effort was made to play down the role of cars in this. RACWA claimed that while it had been established that motor vehicles were responsible for a share of air pollution, it was important to recognize that industry and households also contribute to the problem (RACWA 1970b: 4; 1973a: 8). RACWA considered this kind of pollution 'simply a by-product of economic prosperity' (RACWA 1970b: 4). It was claimed that America had become 'hypnotised with the problem of pollution' and that this had resulted from 'propaganda by environmentalists' (RACWA 1970a: 29). The article referred to the opinions of 'industry scientists' to back up the club's claim that air pollution was not really a very serious problem. One such, unnamed, scientist was reported to have said that 'it is sheer vanity in mankind to pretend that merely human efforts can seriously upset the balance of nature' (ibid.: 29). Environmentalists' efforts to draw attention to the problem were dismissed as 'rabble-rousing' (ibid.: 29).

By the 1990s RACWA had softened its approach on the environment, reinventing itself as an organization with an environmental agenda. An editorial noted, 'motor vehicles have the potential to do great damage to the environment. The RACWA is a club for vehicle owners, therefore the RAC must have concern for the environment' (RACWA 1990a: 2). The Club released a strategy and policy statement on 'motoring and the environment' (RACWA 1991a: 8) but it did little to upset the favouring of cars, 'the strategy acknowledges that motor vehicles are a source of environmental concern. However, their contribution to the problem must be kept in perspective. Environmental controls must be balanced with the need to maintain personal mobility, the movement of freight and the needs of industry' (RACWA 1991c: 5). While environmental initiatives became an increasingly important part of the club's public affairs strategy, the Club tended towards the 'technofix' solution, including initiatives such as a pamphlet – 'How to save petrol, save money and help save the environment' (RACWA 1991b; 1992), support for the Department of Environment 'Smoky Vehicles' campaign (RACWA 1995b), a free emissions testing service for members (RACWA 1996; 2000d; 2000e), the RAC's 'air care campaign' which aimed to educate motorists about environmental pollution, largely through the use of billboards (RACWA 2000c; 2000d; 2000e);

and signing up its vehicle fleet to the 'Carbon Neutral' programme (RACWA 2002a) involving offsetting emissions in a variety of ways including tree planting.

On the depletion of non-renewable fossil fuels, the only article during the 1960s, entitled 'World oil reserves still increasing', was extremely optimistic about the future of global supplies. The article claimed that although the best available estimates suggested that, at the rate it was being consumed the world would have enough oil to last just 39 years, this did not mean that 'oil reserves will be exhausted in 39 years, or sooner, should consumption continue to rise' (RAC 1962b: 11). RACWA confidently predicted that new reserves would be discovered at a greater rate than the increase in consumption and that continual improvement in methods of recovery would provide greater yields from existing fields. Fifteen years later, with less optimism, the RACWA stated, 'one potentially insoluble problem which clouds an otherwise fairly bright motoring horizon, and that is – where will the fuel come from to run our cars?' (RAC 1978: 2). RACWA saw two solutions: conservation and fuel alternatives (RACWA 1979: 1). By the turn of the century RACWA promoted investment in public transport as a way to deal with oil shortage and environmental pollution, 'More trains and buses the key to cleaner air' (RACWA 2000b). In 2005 the Club acknowledged that the 'technofix' solution was 'still a long way off' and suggested the 'while we wait for alternative solutions to energy resourcing, support is growing for a bigger, better public transport system for Perth and WA's regional centres' (RACWA 2005/6b).

Sydney: National Roads and Motorists' Association (NRMA)

The NRMA is a large association of road-users, with almost 2 million members (NRMA, 2004). It is the New South Wales equivalent of the RACV and RACWA. The NRMA is a significant actor in transport policy discourse, claiming 'the highest penetration of membership in the world' relative to the number of private motor vehicles on the road (NRMA, 2001: 16). In this case study we focus on the discourse surrounding NRMA's Clean Air 2000 campaign during the 1990s. The campaign provides an example of a more proactive approach – specifically addressing some of the institutional barriers to sustainable transport. The campaign could be considered a critical juncture. However, as the NRMA no longer holds such a progressive position, this tends to confirm the campaign as an 'exception to the rule' (Learmonth, 2001; APT, 2004). We draw on NRMA magazine, *The Open Road*, together with a content analysis of the daily newspaper, *Sydney Morning Herald*.

In 1992, NRMA undertook a survey through '*The Open Road*' to determine the level of environmental concern among its members. The survey received an unprecedented response, with more than 35,000 members responding to the survey and more than 90% indicating they 'wanted NRMA to pay greater attention to environmental issues'. In 1993, NRMA commissioned the Australian Council of Social Services (ACOSS) to conduct a series of workshops. The findings of the *Open Road* survey and the ACOSS study were then used to produce a series of environmental policies, published in 1994, and used to build the foundations for the Clean Air 2000 campaign (McKenzie, 1995a: 12).

The Clean Air campaign was launched in 1995 with the front page – 'Let's Clear the Air: Join NRMA's campaign' (NRMA, 1995). Inside, it was explained that 'NRMA knows about the benefits and disadvantages of cars … we are in a good position to help Sydney address its air quality problems' (ibid.: 15). The Association was, thus, positioned as part of the solution to air pollution, rather than as part of the problem. Rather than seeing a proactive approach to environmental protection as diametrically opposed to past practice, it was positioned as an extension of traditional activities, 'In the same way that NRMA helps motorists with safe and economical car use, NRMA is now working to help motorists use their cars better environmentally. Cleaner motoring and travel practices are being promoted alongside our traditional areas of interest such as road safety and good roads' (McKenzie, 1995: 30).

The Clean Air campaign was structured around two interrelated strategies. The first was advertising (television, radio, and billboard) to raise people's awareness of air quality issues. The second was to educate people about personal choices they could make to improve air quality, and to widen the choices available to them. The campaign was planned as an 'Establishment Phase' (June 1995 – June 1997) followed by a 'Solutions Phase' (June 1997 – June 2000) (NRMA, 1998). It is this second phase that holds most interest for us since it appears to look towards sustainable transport solutions. Early on, however, it was recognized by NRMA that the transport choices individuals could make to improve air quality depended upon the choices available to them:

> improving the quality of alternatives to motor vehicle travel is a difficult problem that requires the involvement of a range of organizations including government, business, industry, science, environment and community groups. NRMA's answer to this problem is to bring together key community leaders and decision makers to implement strategies that improve the opportunities for motorists to reduce their vehicle emissions – the Clean Air 2000 Task Force. (McKenzie, 1995: 30)

The Task Force oversaw the development of the campaign, providing high level direction and guidance through regular meetings and working groups. Organizations represented on the Task Force included: governments (federal, state and local); research organizations (the CSIRO, Sydney University and the Australian National University); corporate stakeholders (the Motor Traders' Association of NSW, Shell Australia, and General Motors Holden Australia); and non-government organizations (Total Environment Centre). Of note is the absence of the NSW Roads and Traffic Authority which could have been expected to have a strong interest in the initiative. Other members of the Task Force reinforced the Clean Air campaign's stance that air pollution is a problem which belongs to the whole community: for example, the Australian swimming team, Radio Triple J, the Sydney Symphony Orchestra and Telstra (NRMA, 1998). The task force was chaired by Rod McGeoch, who had recently won the 2000 Olympic Games for Sydney, and therefore had a personal interest in delivering on the claim to 'clean air', which was a significant factor in Sydney's winning bid (McGeoch, 1995: 14).

The 'Clean Air 2000 Strategic Action Plan', was released in 1998 with the claim that 'the community is becoming increasingly frustrated with the lack of genuine progress towards the development of appropriate solutions'. It proposed the setting of 'community targets' which 'must be realistic, but not conservative … designed to test the upper limits of achievement'. It was noted that achievement of the targets would not only require community support and action, but also institutional reform of government decision making processes to demonstrate a high level of government consideration of air quality objectives in transport planning and budgetary decisions (NRMA, 1998). Four key objectives were set:

- 'Improve Sydney's urban public transport system, particularly in Western Sydney' (to reduce 'car-based commuting').
- 'Improve the emissions performance of the Sydney vehicle fleet'.
- 'Improve cycling and walking networks across the Sydney metropolitan region' (to encourage 'reduced car usage for short trips').
- 'Mobilise all levels of the community (government, business, industry, public) towards the adoption of strategies and actions that reduce urban air pollution and traffic congestion' (success will be dependent upon 'all sectors of the community' taking 'ownership' of pollution reduction strategies). (NRMA, 1998)

Measurable targets were set for each of these objectives. They were ambitious, for example, 'Not less than 70% of people who travel [should be] using public transport at least once a week by the year 2000' (NRMA, 1998).

Improving the range of transport choices available to individuals was a central element of the Clean Air campaign; this in stark contrast to the discourse of RACV and RACWA which continued to argue for car use, even within the public transport discourse. McKenzie described improving 'the opportunities for individuals to reduce vehicle emissions' as an essential part of any strategy to reduce vehicle emissions (McKenzie, 1995: 29). An article published in the *Sydney Morning Herald* in May 2000 would seem to suggest that the storyline of 'choice' remained alive in transport debates, 'Commuters would choose public transport … if they had a genuine choice', later elaborated, 'What on the surface might seem an expression of choice [Sydney's current high levels of car use and correspondingly low public transport use] is really the outcome of an absence of genuine choice' (Walton, 2000: 13).

The setting of a target to use public transport one day a week was an interesting contrast to the usual storyline that 'Private cars are essential to modern life'. The campaign, however, unequivocally accepted that private 'cars form an integral part of our society'. Importantly, this acceptance was less unconditional than in the past, 'Through the adoption of a more holistic approach to road based transport, NRMA will show that the environmental impacts of road travel can be managed by using our cars more wisely and perhaps a little less when viable alternatives are available' (McKenzie, 1995: 30).

'Balanced Transport' was a core storyline during the campaign. The NRMA argued that only a shift away from excessive use of the private car and towards other modes of transport (public or non-motorised modes) could merit the description 'balanced':

> [NRMA believes] that a balance between major new road projects and changes in the public's travel behaviour will be needed to make the best use of the existing road network in urban areas … NRMA sees the New Southern Railway (and rail improvements generally) as an opportunity to balance modal share and divert users of private road transport to rail. (NRMA, 1995a: 5)

In the discussion paper *Shaping Sydney's Transport – a framework for reform*, the Clean Air 2000 Project Team identified 'an apparent and urgent need to ensure transport planning and provision is carried out in a far more balanced fashion than it is now' and suggested 'some reform of current transport planning, funding and provision practices and arrangements' (NRMA, 1996: 16). This spin on the 'out of balance' storyline was repeatedly taken up by the *Sydney Morning Herald* (Beale, 1996b; Hogarth, 1998a; Hogarth, 1998c; Hogarth, 1998e; Wainwright, 1998; Wainwright & Totaro, 1998; and Walton, 2000). For example,

> The NRMA-led Clean Air 2000 Task Force has called for immediate action to address Sydney's future transport needs in suburbs beyond Parramatta … Sydney has become a city divided, with vast areas of its western and south-western fringes poorly serviced by public transport and likely to suffer worsening traffic and deteriorating quality, the task force says in a report … Task force representative and the director of the Total Environment Centre, Mr Jeff Angel, warned that Sydney had less than five years to correct its transport imbalance and 'air quality apartheid'. (Morris, 1996: 4)

There was also a clear shift in the discourse 'Road construction is necessary for economic health' to an emphasis on the total costs of car use, '[Vehicle emissions] degrade the environment and affect our health … In the end, we have to weigh the benefits and disadvantages of travelling by car versus other options. Cars certainly make our lives easier, but there are environmental, social, health and economic gains to be had from driving less frequently' (NRMA, 1995: 16). The environment was seen as a competitive advantage, 'Timely action [with respect to the protection of air quality] will also assist in the management of the city's long term economic prosperity' (NRMA, 1996: 8).

What is interesting is that such discourse has been used in the anti-road discourse for decades. The reasons for NRMA to adopt such a discourse may be the result of including a wider range of representatives on the Task Force than was normal for the NRMA. Another factor may be that increasing environmental consciousness during the late 1990s forced pro-road advocates also to acknowledge this. Alternatively, as it became apparent just how poor the transport choices were and how damaging the effects of a dominant focus on the private car had been, the storylines for a roads based transport policy were finally exposed for what they were – stories.

The Clean Air campaign appears to have been generally well received by the public. Of the 41 references to the campaign in the *Sydney Morning Herald* archive, we scanned 21. The single exception to the positive coverage appeared in the 'Motoring' section, in an article written by Tony Selmes, the executive director of the Motor Traders' Association of NSW (Selmes, 1995). Occasional references continued to be made to the campaign even after it was abandoned in late 1998. During 1996 some measurable progress in terms of community behaviour change (buying smaller cars, getting rid of a car, working from home, joining a car pool, and increased use of public transport) was reported and suggested to be a result of the campaign (Beale, 1996a: 11; and Beale, 1996b: 5). When the Task Force released its detailed proposals, these received front-page coverage under the heading 'How to squeeze the cars out of Sydney: $5 billion solution to transport chaos':

> A radical $3.5 billion to $5 billion masterplan has emerged to cure Sydney's transport ills and help the State Government deliver its pledge to slash the use of private cars and push commuters onto public transport. The key proposal of the three-year project for the NRMA's Clean Air 2000 task force – which has yet to be presented to the State Government – calls for a mix of public and private investment in rail, bus and light rail. The overall blueprint aims to fill the public transport gaps in the north-west, west and south-west, and northern corridors, bringing an extra 200,000 commuters daily into and around the greater metropolitan area... NRMA staff developed the masterplan by reassessing all proposals over the past 25 years by governments, councils and lobby groups. (Wainwright & Totaro, 1998: 1)

In late 1998, despite an overwhelmingly positive public response, the campaign folded. An article entitled 'Haze hangs over future of Clean Air 2000' appeared. Three weeks later, in an article entitled 'Transport revolution urged to fight smog', Hogarth referred to the campaign as 'the NRMA's now-defunct Clean Air 2000 initiative' (Hogarth, 1998c: 5). Evidently the required financial support for the initiative was not found.

The discursive shifts achieved during the Clean Air campaign were very significant, especially in Sydney where there had been a longstanding 'lack of genuine progress towards the development of appropriate solutions' (NRMA, 1998). It is probable that NRMA's advocacy of (and achievement of) institutional shifts during the campaign – most noticeably through bringing together a range of organizational actors, essentially an advocacy coalition (Sabatier, 1998), through the Clean Air Task Force – assisted community uptake of its modified storylines; and actors who 'walk their talk' do tend to be more credible than ones who do not. This suggests the importance of an integrated approach to barriers to sustainable transport.

Stakeholders: Pro-Public Transport

The pro-public transport and anti-roads groups, unlike the pro-roads groups, do not operate within a larger umbrella organization, and they vary in the three cities. Melbourne's Public Transport Users Association is the largest in membership and longevity and the strongest on conceptual arguments. There is no equivalent organization in either Sydney or Perth. Instead we have examined the discourse of two groups, each established initially to fight a road proposal but which have turned to the public transport solution. These stakeholder groups focus mainly on the environmental storylines, and the social virtues of public transport, but have also countered the pro-roads storylines directly. Content analysis is made of each group's media releases, newsletters and websites.

Melbourne: Public Transport Users Association (PTUA)

The PTUA was founded in 1976 as the Train Travellers Association. It is 'the recognized consumer organization representing passengers of all forms of public transport' (PTUA, 2001). In its publications, the PTUA both identifies, and positions itself in opposition to, many of the storylines present in the pro-roads discourse. This joint manoeuvre of identification and positioning is present throughout all PTUA publications but it appears most explicitly in the section headed 'Fallacies from the road lobby' (PTUA, 1995: 16–26) where the following pro-roads storylines are systematically identified and challenged: 'chronic congestion versus free movement', 'infrastructure investment causes economic growth' (and the associated 'freight' story), 'low urban density can only be served by roads', 'sustainability' / pollution, 'bypass', and 'linking'. The PTUA uses its own storylines to present an alternative to the transport scenario offered by the road lobby – a first-class, international best practice public transport system can provide a real alternative to the mobility offered by the private car, without the heavy environmental, social and financial costs imposed by car-dependency.

PTUA rejects the road lobby storyline by arguing that the 'costs of freeways far outweigh the benefits' (PTUA, 1991: 4–5; PTUA, 1995: 5–6; PTUA, 1997: 5–7). The social, environmental and economic components are listed: social costs – severance, social inequity, public health problems, noise, vibration, 'dead' streets (cycling and walking discouraged by noise and smell of cars as well as increasing distances needing to be covered), greater proportion of time spent travelling; environmental costs – inefficient land use, urban sprawl, greenhouse emissions, smog, polluted run-off, economic costs – costs imposed on individuals by the 'necessity' of owning a car, cost of road accidents, cost of damage to buildings by air pollution, increased trade deficit (oil imports). The PTUA argues that, 'Freeways are unnecessary' because the passenger task of radial freeways would be more economically achieved, with less social and environmental damage, by excellent public transport, leaving existing radial roads much freer for freight, commercial and cross town travel (PTUA, 1995: 16–20; PTUA, 1997: 9–12). PTUA asserts that 'Freeways are an outdated solution to the transport problem' (PTUA, 1995: 30) and instead that 'Livable cities require excellent public transport' (PTUA, 1995: 5–6, 30; PTUA, 1997: 5). Excellent public transport is defined as follows: 'First class service is key: integrated, frequent, clean, reliable, safe, fast, and affordable public transport increases patronage, decreasing public transport deficit (PTUA, 1991: 15, 19–25, 29–30; PTUA, 1995: 26–30; PTUA, 1997: 24–26, 28). Here, interestingly, the PTUA chose to engage with the 'deficit' storyline rather than counterpose a different a storyline.

The PTUA played a role in positioning itself with respect to other actors in transport policy, particularly in its submissions on policy documents, where it drew on a detailed analysis drawing from international literature on (and experience in) transport planning. For example,

> The Public Transport Corporation is the single most important player in transport planning. The PTC can help by adopting and implementing the ideas of this document[2] and by continuing its current efforts to become more reliable and customer-oriented.
>
> VicRoads has commendably re-thought its approach to transport planning over the last few years, moving away from the idea that transport problems must always be solved by building more roads. VicRoads must continue with this approach in searching for total solutions to transport problems, by:
>
> - Working closely with the PTC to provide optimal operating conditions for road-based public transport;
> - Not building arterial roads which compete with public transport;
> - Helping councils introduce area-wide traffic calming measures. (PTUA, 1991: 33)

Whilst making the case against the road lobby, PTUA argue against the 'predict and provide' storyline, 'Trend planning … unquestioningly takes the current trends and perpetuates them. [Instead] Planning should set aims and objectives and show how they can be achieved' (PTUA, 1991: 6). The PTUA makes some attempt to explain why it is so difficult to effect a change in transport policy in Melbourne – the barriers to sustainable transport. It points to the 'dominance of an 'engineering mentality':

2 Referring to the policy document 'Greening Melbourne with Public Transport' (1991).

> many proposals for public transport take their cue from the "engineering mentality" of the road planners and try to 'fight infrastructure with infrastructure'. So we see "imaginative" ideas for monorails, levitating trains, rebuilding the outer circle railway and so on. This is not transport planning .. Transport planning concentrates on people – the people who need to be persuaded to use public transport. It asks what their needs are before it asks what technologies are available, or how they can be fooled into liking public transport. (PTUA, 1995: 24)

PTUA also suggests that there is an inability to believe that good public transport is possible, 'because this [high quality public transport] is something that has never been seen in Melbourne, many people have difficulty imagining it, or believing that it is possible, but in overseas cities with well-run public transport it is regarded as perfectly normal' (PTUA, 1995: 27). Finally, PTUA point to the inadequacy of community input during the planning process:

> In stark contrast with the practice in Melbourne under both the current and previous governments, [Vancouver's Livable Region strategy] was developed after three years of extensive consultation in which the community was given the opportunity to set goals for the city. (PTUA, 1997: 20)

Sydney: EcoTransit

Three groups have all opposed freeway building in Sydney: Action for Public Transport (APT), EcoTransit, and Truth About Motorways (TAM). EcoTransit appears to be the closest operational equivalent to the PTUA, running its own campaigns which promote positive alternatives to continued freeway building. For this reason we focus the Sydney case study on EcoTransit's role in transport discourse.

EcoTransit is the reincarnation of Transport Action Group Against Motorways (TAGAM), which itself originally branched off from Wolli Creek Preservation Society (WCPS) in 1993. TAGAM became EcoTransit Sydney in 1999. This name-change reflected the change in tactics adopted by the group – in order to stop freeways and tollways, public transport proposals had to be advocated in their place. The 'eco-' prefix stands for 'eco-nomic' as well as 'eco-logical': 'The less resources used by the transport sector, the more efficient our economy is and the less damage is done to the environment' (EcoTransit, 2001a). Like the PTUA, EcoTransit had a critical mass of people with technical and professional knowledge. Our content analysis includes the EcoTransit website and 14 articles from *Hell on Wheels* (TAGAM/EcoTransit's occasional newsletter) over the period 1993–2001.

Like the PTUA, EcoTransit rejects road lobby storylines, counterposing the storyline that the 'costs of freeways far outweigh the benefits' (with many examples of different kinds of cost), arguing instead that public transport offers a better solution:

> Urban freeway building is counter-productive and self-perpetuating. As soon as a new freeway is opened, traffic volumes quickly grow to fill the available road space. Bottlenecks and congested choke-points that the original freeway was intended to remove, are simply moved to another part in the network and the process is repeated ... The M6 is not a 'solution' to the transport crisis. Removing vehicles from congestion hot spots by providing alternatives to the car is the real solution. The Bay Light Express and other public transport upgrades (that are long overdue) do hold out the prospect of real solutions. (EcoTransit, 2001b)

EcoTransit explicitly rejects the 'Roads are necessary for economic prosperity' storyline used by the road lobby, 'The RTA line is that if a city doesn't continue to build bigger roads, the economy will grind to a halt. This is nonsense. In fact evidence shows that cities which place greater emphasis on public transport development and less on motorways, are more efficient and more productive' (Zeibots, 1997).

Over the whole period EcoTransit has argued for high quality public transport as an alternative solution for transport policy. In 1993 the group stated that the 'RTA's plans for the region ... are barbaric and outdated. Rail is the way to go' (EcoTransit, 1993a). They used evidence from public inquiries which support their case:

> There have been no less than three commissions of inquiry into Sydney freeways – the F1, F5 and F2. All three inquiries concluded the freeways should not be built and that public transport and rail freight development should be given priority. (EcoTransit, 2001b)

Again like the PTUA, EcoTransit proposes high quality public transport (and better facilities for cycling and walking) as the alternative to continual motorway construction (EcoTransit, 2001d).

A particularly strong counterposition to the 'predict and provide' storyline in the EcoTransit literature is 'untrustworthiness', 'Eight years of intimate collaboration between the RTA and private tollway operators has ended with the RTA engineers becoming spindoctors for the tollway consortia. In terms of honest and reliable advice to government and the community the authority's integrity can't be relied upon' (Zhukov, 1997). Such strong claims were based on considerable research and analysis. For example, in an article entitled 'Inventing a fraudulent present to disguise an unpopular future', EcoTransit exposes RTA's future traffic modelling as extremely flawed by showing that 1994 predictions for the year 1996 were proved, in hindsight, to be up to 62 percent above actual measured figures in 1996. 'Untrustworthiness' is also raised by Melbourne's PTUA (PTUA, 1995: 16–17) and Perth's TRAC (below) about the respective State authorities.

The storyline 'pitfalls of private sector financing' forms a strong focus in EcoTransit documents, it is also present as an undercurrent in PTUA literature. It links to the above 'untrustworthiness' storyline and to the Economist's Tale:

> There is compelling evidence that the RTA has deliberately exaggerated key traffic forecasts for the Eastern Distributor. These forecasts support an impression that tollable traffic might be sufficient to service debt on the $650 million private project. Similar forecasts have appeared in the prospectus for the public float of the Macquarie Bank-sponsored Infrastructure Trust of Australia. Did the RTA advise the Minister for Roads that the potential revenue stream of the Eastern Distributor was sufficient for the operator to service debt? If they didn't, what was the basis of the government's selection of the preferred tenderer, its negotiations with the preferred tenderer, its decisions on design modifications?
>
> The implications should not be lost on the government. The financial collapse of a motorway built with funds raised by a stock market float would drag down with it hundreds of millions of dollars invested by institutional investors, mainly superannuation funds. In this, the recent history of Labor governments in Victoria, South Australia and Western Australia should be flashing warning signals to the Carr [NSW Premier Bob Carr] Government. (EcoTransit, 1997)

Perth: Transport Action Coalition (TRAC)

In Perth we turn our focus to TRAC – a community group established to oppose the construction of one road, the Fremantle Eastern Bypass (FEB). Although in existence for some years before, TRAC adopted a constitution in March 1998 and became a legal entity. The group did this in order to accept members and so raise funds to fight the Bypass and establish a legally recognized body of people who could challenge the proposal (TRAC, 1998). TRAC expanded its interests as the transport discourse unfolded, taking on a wider view than simply the FEB as they recognized it as part of a 'motorways by stealth' argument (found in the 'missing links' storyline) and eventually came to challenge all road proposals in the sub-region, and new car park proposals (which would generate car trips). TRAC set out to 'seek environmentally sustainable solutions to regional traffic problems' (TRAC, October 2000). The group also initiated FRETRAN, (Fremantle Regional Transport Network) which consisted of 12 community groups opposed to the Bypass and dedicated to finding different solutions to traffic problems (TRAC, 1997b).

We conducted a content analysis of TRAC's media releases (1997–2005) and Newsletters (1997–2003). The core storylines were 'motorways by stealth', the social impacts of roads (community severance), the environmental impacts (noise, pollution/air quality, emissions), the 'freight' storyline. Like PTUA, they were continually concerned with the planning process and its failings in relation to community input. They argued for alternatives to road building, including making better use of existing infrastructure and public transport.

TRAC positioned itself to counter the storyline 'bypass and bottleneck', counterposing this with their storyline 'motorways by stealth'. It was noted by TRAC that the 'Bypass is not merely a 3 kilometre stretch of road but part of the very large and significant Western Suburbs Highway' (TRAC, 1997), this despite the fact that 'that Main Roads WA had successfully argued [to the Environmental

Protection Agency] that the bypass was just a little, insignificant isolated road' (TRAC, 1998a). TRAC pointed out that, 'Main Roads' claims that the bypass will relieve local traffic problems were disproved by their own traffic studies' (TRAC, 1998b). They also used the 'bypass' storyline to highlight that it would simply create a new bottleneck and instead better use of existing roads should be made:

> Building the bypass and the CAH[3] will ensure that congestion and a bottleneck will occur on the Stirling Bridge in the near future' (ibid). TRAC argued that instead of building a new road (FEB) there should be traffic calming of Fremantle's local road – Hampton Road – which the local council implemented, and that an existing major arterial road could be more efficiently used to take any predicted growth in traffic. (ibid.)

TRAC also used public display as a way of communicating its discourse, appearing each year at the annual Fremantle Festival Parade, in one year turning the TRAC logo into a 3D model of a heart, 'representing the heart of the community broken by a road, floating down the streets on the shoulders of volunteers' (TRAC, 1998a). This is a direct contrast to the way in which Melbourne's City Link Authority used the heart image in the early discourse (see Figure 3.1 Chapter 3). TRAC noted, 'It was heartwarming to hear the shouts of support and approval as we walked through the huge crowd gathered. The judges were also affected by this image and chose this float as the winner of the most environmentally sound float in the parade' (ibid.).

The opportunity for public input in the decision was at first denied when the Environmental Protection Authority (EPA) made a decision not to assess the FEB. TRAC joined forces with other community groups and the City of Fremantle to lodge a successful appeal against the EPA decision in February 1997. TRAC turned to the 'sustainability' storyline to counter the arguments for the FEB pointing to the social concerns of community severance and 'the environmental impacts on remnant vegetation and conservation areas in the highway's path' and 'the highways potential contribution to Perth's smog problem and greenhouse gas emissions' (TRAC, 1998a). Although successful, the EPA afforded it only a low level of assessment[4] and TRAC appealed a second time, arguing that the social impacts of the highway and the whole route must be assessed, not just the

3 CAH – Controlled Access Highway.

4 Under the Western Australian Environmental Protection Act 1986 – Environmental Impact Assessment (Part iv Division 1) Administrative Procedures 2002 the five levels of assessment are:

(a) Assessment on Referral Information (ARI); (b) Proposal Unlikely to be Environmentally Acceptable (PUEA);

(c) Environmental Protection Statement (EPS); (d) Public Environmental Review (PER); and

(e) Environmental Review and Management Programme (ERMP).

FEB (TRAC, 1998b). While the appeal succeeded and the EPA raised the level of assessment to 'PER', TRAC noted that the assessment would be quite limited:

> The Minister, Cheryl Edwardes, went into some detail about the EPA's ability to carry out social impact assessments. The effect was to limit the areas the EPA could assess the proposal on social impact grounds [sic]. She rejected an assessment of issues such as traffic noise, vehicle emissions, light overspill from street lighting and visual amenity suggesting that MRWA was able to address such matters. However, the EPA is to assess the effects of the highway proposal on the potential social values of Clontarf Hill due to the disturbance of vegetation. (ibid.)

TRAC expanded its scope to take on a wider role in the transport debate. It commented on other major redevelopment proposals focussing on the traffic generated by them and turning to the failure to look at transport alternatives. On the Fremantle Waterfront Redevelopment Plan, for example, 'the current proposal will create major traffic problems in the West End. The plan proposes to create an extra 900 car parking bays around Victoria Quay, and gives little scope for extra and improved public transport. This gives people no alternative but to continue using their private vehicles to access the area' (TRAC, 1998b). TRAC also drew on international ideas as part of their sustainability discourse, 'In 1997 Friends of the Earth in England, among others, were able to convince the British Government to introduce a bill which would encourage the reduction of traffic volumes through the setting of achievable and enforceable targets. The aim of the targets is to reduce the adverse environmental, social and economic impacts of road traffic' (TRAC, 1997a).

When Government argued that the FEB was needed to get freight off other major roads, TRAC actively engaged to counter this 'freight' storyline, arguing for an evidence base to the bureaucrats stance, '... there still needs to be a study of truck movements and how much of that can be transferred to the rail system' (TRAC, 1999). TRAC demanded that the government, 'stop subsidising the road freight industry and channel that money into developing an effective alternative freight transport system' (ibid), arguing for better use of existing roads. TRAC noted that the federal Goods and Services Tax had reduced the price of diesel fuel to the road transport industry and argued, drawing on key research studies, that this would have major environmental and health impacts but that, 'This evidence is ignored in the push to give further subsidies to the road freight industry. Recent studies in Australia indicate that the road freight industry receives hundreds of millions of dollars in subsidies and causes billions of dollars of negative impacts' (ibid.).

Finally in October 2000 TRAC announced a party to 'celebrate the end of the Fremantle Eastern "Bypass"' and outlined their campaign 'for a sustainable transport system' where other road reserves would be deleted and the land sold, 'road freight pays appropriate road user charges that reflect road damage and externalities. Improve the proportion of freight moved by rail ... Main Roads WA objectives widened to include provision of light/heavy rail routes, greater emphasis on cycleways, bus lanes, pedestrian routes', and new rail links be provided (TRAC, October 2000).

Conclusion

We have demonstrated the strength of the pro-roads groups. They operate as part of a 'discourse network', sharing a common argumentative language with their respective State government roads agencies (see Low, 2005). There was, therefore, no need for specific advocacy *coalitions* because the case for road building formed the core of the shared language. Analysis of their documents shows that the Clubs were constantly on the offensive: identifying and refuting opposing arguments – a nimble footed advocacy employing a variety of specious and circular propositions, appealing to emotional arguments but condemning as irrational the emotional arguments of their opponents. The exception to this *modus operandi* was when the NRMA did, innovatively, form an advocacy coalition bringing in actors outside the roads lobby, with the result that the common language was severely disrupted

In contrast, the pro-public transport groups operated outside government and used a language contrasting with the conventional wisdom of roads; each can be considered a lobby group. They employed two discursive strategies: one counterposing their own storylines against the 'path', and the other engaging with and refuting the 'path' storylines. Perth's TRAC achieved some success; again community opposition was effective in gaining political support (see Chapter 9) to remove a section of freeway. Melbourne's PTUA and Sydney's EcoTransit had less success in stopping freeways.

Chapter 11
Contemporary Mental Models

In this chapter we draw the empirical research to a conclusion by turning to the way in which those involved in transport policy and planning construe transport problems and solutions. We investigated what we call their assumptive worlds or 'mental models' We examined the extent to which the storylines extracted from published policy documents (Chapters 8 and 9) actually inhabited the minds of policy makers. We achieved this by conducting an interview survey with key respondents in each of the three cities. We termed these respondents 'policy shapers', and they included those who had, or continued to have, significant influence in transport policy in each state. These included State politicians, senior public servants and executive staff, professional consultants, and representatives from transport stakeholder groups. What will be shown is that, overall, there is a high level of agreement with the most influential storylines. In some cases, though, the key premises are challenged and we start to see the emergence of support for the new counter positional storylines sitting within the Sustainability Tale. There is also some evidence that particular sets of storylines correlate with the 'functional' group to which the key informant belongs. This is most apparent for those from the roads-based group where those storylines rooted in the 'Engineer's Tale' continue to form strong mental models for them.

Our research approach comprised a face-to-face interview with each respondent of between 60 to 90 minutes duration. We employed a questionnaire survey instrument comprising 71 statements (Appendix 3) to which respondents were asked to record their agreement or disagreement using a 7-point Likert scale. The same survey instrument was used in each city with only limited adaptation made for differences in nomenclature (for example 'freeway' in Perth rather than 'motorway' in Sydney).The majority of statements were taken from policy documents and annual reports, and represented lines of argument that were used to justify investment in transport infrastructure, in particular to rationalize the balance between investment in road and public transport infrastructure. Some statements were included to represent alternative storylines that could provide a more sustainable approach (see for example Q7a in contrast to Q7b). The set of storylines for the Sustainability Tale was also enlarged in order to gather information on the extent to which they were accepted by the policy shapers. In practice the 71 statements also had the effect of eliciting positive or negative comments on the storyline contained in the question and its underlying assumptions. This part of the research generated a large volume of data. In this chapter, due to space limitations, we report the overview of the findings rather than the detail of qualitative data that accompanied each response.

A purposive sampling technique was employed (Neuman, 2000). People were selected based on their ability to influence policy in their respective organizations or as consultants. People either had worked, or were continuing to work, on transport policy documents and/or in transport policy-making positions. Despite the political sensitivity of the subject we achieved a reasonable representation from those who had influence over roads policy and public transport policy (subject to the qualifications mentioned below). The group of respondents is not a statistically significant representative sample. Such a sample would be hard to define given the difficulty of defining the boundaries of the population of policy shapers. Our choice of respondents was also restricted either by their availability for interview, or the organizations' willingness to allow people to participate. A particular account of who was interviewed is given for each city below. For reasons of confidentiality the interviewees cannot be identified.

In Melbourne the interviews were conducted during 2003. Twenty people were interviewed. They included: five State politicians, five senior public servants, four senior executive staff of local councils, four representatives of transport lobby groups (also consultants), and two senior transport consultants. Unfortunately the road planning authority, VicRoads, chose not to allow its staff to be interviewed. Nevertheless as wide a range as possible of other groups were represented and at least one respondent was formerly a senior executive of VicRoads.

At the time of our interviews in Sydney (late 2006), preparations were under way for an election. This limited our access to State politicians. We were unable to gain access to State Ministers or policy advisors. It proved very difficult to obtain interviews with the Ministry of Transport and the State Transit Authority. They are not directly represented by current employees, but the respondent group includes previous employees of these organizations. Eventually, 13 people agreed to participate. They included: eight senior public servants, two transport lobbyists (also consultants), and three senior transport consultants. During the interview period Sydney released its *Urban Transport Statement* regarding strategy for the delivery of transport infrastructure, and this may have influenced the views of the respondents due to its immediacy in their minds.

In Perth, unlike Sydney or Melbourne, there was no difficulty with willingness of the policy shapers to participate. This simple fact suggests a greater openness (and trust) on the part of the respondents. This in itself may provide some insight into why there appear to be fewer institutional barriers to the progress of sustainability in Perth compared with the other two cities. The interviews were conducted during 2006. Twelve people agreed to be interviewed. They included: one State politician, four senior public servants, one senior executive representing local government, five transport lobbyists (representing pro-roads/freight, anti-roads, pro-public transport) and one senior transport consultant.

Comparing the Mental Models of Policy Shapers in each City

The Engineer's Tale

Taking the Engineer's Tale first, of the 14 different storylines presented, 10 received a mean score higher than 4 (where scores above 4 range from agree slightly (5), agree (6) and agree strongly (7) (see Figure 11.1). It is evident that all three questions capturing the 'predict and provide' storyline continue to resonate with respondents. Respondents in Sydney and Melbourne stand apart from those in Perth in terms of their acceptance of the classic predict and provide discourse (Q1 and Q2 – both concerning use of demand projections and modelling). In Perth the respondent group was less inclined to support this approach.

Policy shapers in all three cities shared the belief that it was important to provide for high levels of mobility. Three questions focused on the 'free movement' storyline (Q4–Q6) and here the response was quite nuanced. In relation to question 4, attaining a transport system where people could move freely and quickly, the mean score for respondents from all three cities was 'slightly agreed'. However, it would appear that, in the new century, policy shapers had come to accept that traffic congestion was inevitable (Q5). In a marked contrast to the rise of the freight storyline in policy documents towards the end of last century, free movement for freight in terms of travel time (Q6) was one of the least supported storylines in the Engineer's Tale (although less so for Melbourne respondents – but the three year separation between interview periods may have also had an effect). Looking more specifically at storylines for resolving traffic congestion – 'bypasses' (Q12) and 'bottlenecks' (Q13) – respondents from Sydney and Melbourne were slightly more in accord than those in Perth with the 'bypasses' solution, and held stronger beliefs that 'bottlenecks' must be relieved in order to resolve congestion. It would appear, therefore, that when respondents are presented with more concrete examples of the 'free movement' storyline they continue to support them. This can also be seen in relation to the 'missing links' (Q15) storyline.

Two questions explored the 'balanced transport storyline – Q7a on balanced investment between each of the modes and Q7b on equal balance between 'only roads and 'all other modes'. The former was the most strongly agreed storyline by respondents in all three cities (slightly higher in Perth). The latter was disagreed with. Again this presents quite a marked contrast to what has actually occurred in each city (where funding and implementation of road infrastructure projects has taken the lion's share), and in published policy (where the modes have not been treated equally). It raises questions about the extent to which policy shapers in general are able to assert influence to shift the path.

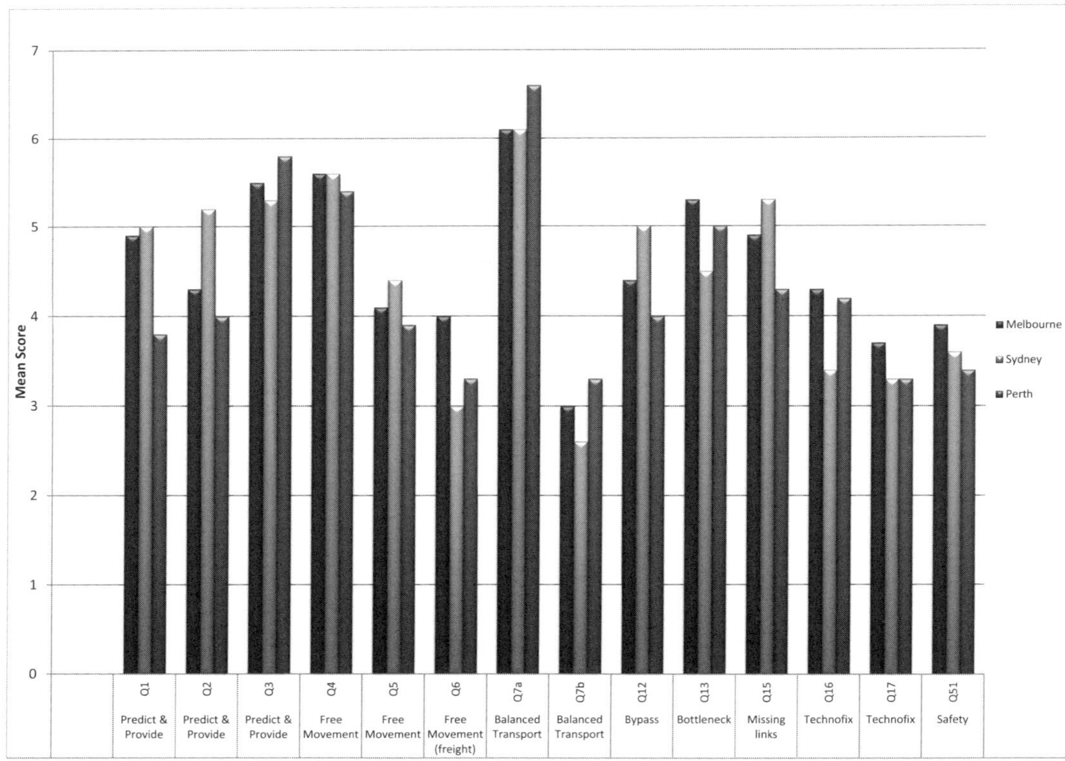

Figure 11.1 Mental models: The engineer's tale

Two questions focussed on the 'technofix' storyline which provided a rationale for continued road building – creation of environmentally friendly vehicles (Q16) and new road use technologies (road pricing for example) (Q17). Neither storyline received a strong mean agreement, although Q16 did receive a slightly higher score in Perth and Melbourne. This may be a sign of the potential for this as a strengthening storyline for road building – and indeed this is evident in the pro-roads stakeholder analysis (Chapter 10). For the 'road safety' storyline (Q51), policy shapers did not award it a high mean score in any of the cities whereas the pro-road stakeholder groups relied on this storyline over many years as a rationale for investing in new road building (Chapter 10).

Overall there is considerable consensus in the response from policy shapers in all three cities to the Engineer's Tale. Small differences can, however, be detected. Perth's 'policy shapers' were found to be in stronger agreement with those storylines that represented a more tempered approach and also had lower levels of agreement than the other two cities on many of the entrenched storylines reflecting mid-twentieth-century thinking.

The Economist's Tale

The most strongly supported storylines in the published policy discourse around public transport were framed under the Economist's Tale. These were the 'public transport deficit' (Q54) and 'public transport as a welfare option' (Q32). Yet what we see in the new century is little support for either storyline, with Perth interviewees showing the lowest mean score for the deficit storyline (Figure 11.2).

However, five storylines relating to lifestyle choice and opportunity as reasons for the continued need for roads (Q43, Q18, Q25, Q44, Q45) continued to resonate more strongly. Of these five, 'travel as opportunity' (Q18) gained a high mean score in all three, with Perth interviewees showing the highest mean score. Next to this, the storyline 'people's choice', with the statement that road building was a response to public demand (Q44), received a high mean score for Sydney interviewees and even higher for Melbourne.

Three statements explored the 'market solutions best' storyline (Q20–22). Of these the statement that 'Congestion pricing is an efficient way of constraining travel demand to fit a given level of transport infrastructure provision' received the highest level of agreement. This is mirrored in all three cities. This again reflects a position where the policy shapers' mental models have evolved compared with much of published policy to a point where it is now seen as acceptable to manage demand. This position is supported too by a reasonable level of agreement that congestion pricing is a fair way of constraining travel demand.

One statement was employed to explore views about a role for private companies in transport operations (Q19) – this received little support.

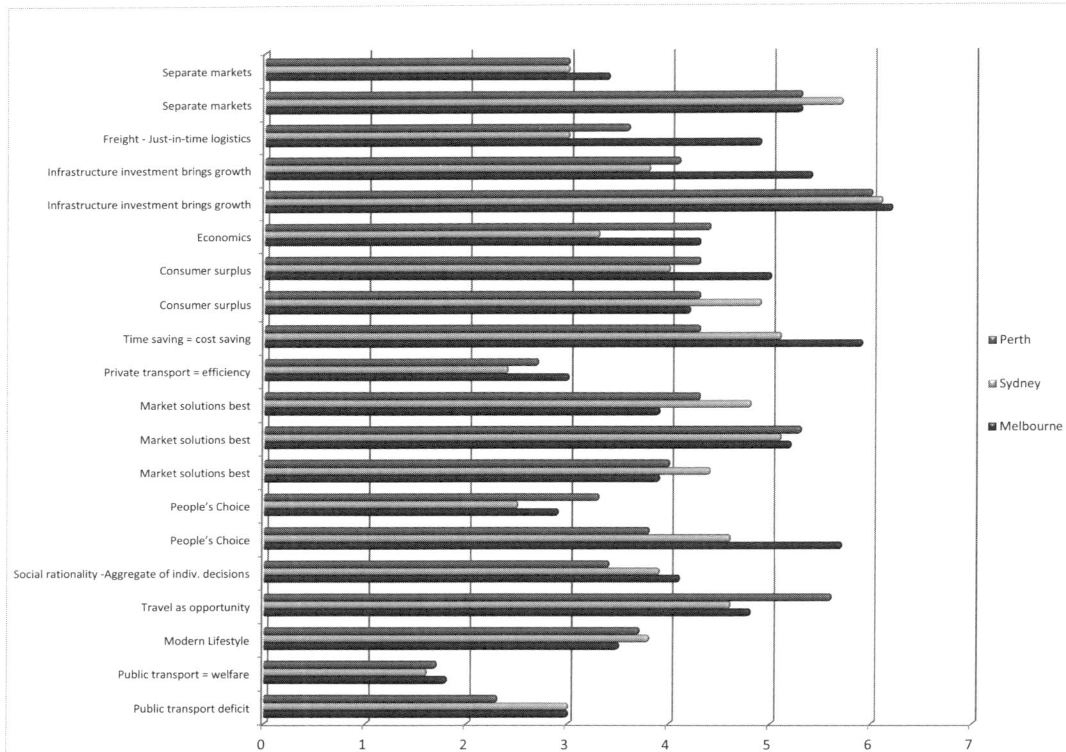

Figure 11.2 Mental models: The economist's tale

The costs and benefits of travel and infrastructure projects were explored using three statements (Q14, Q23, Q24). For the Perth policy shapers each of these statements achieved the same mean score (agree). For Melbourne's policy shapers, there was stronger agreement that 'time saved in travel can reasonably be equated with cost savings for firms' (Q14), slightly less so for Sydney interviewees. This, nevertheless, demonstrates the way in which conventional cost-benefit assessment tools continue to play a strong role in transport infrastructure decision making. What is interesting though is that when turning to consider whether 'limiting or constraining car access will have a severe impact on the local economy (local retail/business)' there was much less agreement.

The belief that 'investment in all the main types of transport infrastructure (roads, railways, airports) contributes to growth of employment opportunities' (Q26) received the strongest level of agreement of all the storylines for the Economist's Tale in all three cities. A more detailed set of statements aimed to break this down by type of infrastructure, with interviewees being asked to score each item out of 20 (see Figure 11.3). Transport interchange facilities achieved a high mean score in all three cities, followed by road infrastructure then rail. There were interesting contrasts between the cities, with Perth interviewees scoring modes alternative to car (cycling, walking, light rail) more highly than the other two cities. Sydney interviewees expressed a strong score for heavy fixed rail and a slightly lower score for bus based infrastructure. This is not surprising since the heavy rail network is so crucial to Sydney's commuter travel.

Turning to the statement which was designed to articulate the freight storyline (Q29), which has been gaining strength in the policy discourse over recent years, we see a mixed response from the policy shapers. Melbourne interviewees showed the highest mean score for the statement 'today's logistical methods in industry, based on the principle of "just-in-time" delivery, mean that more road investment is needed to meet increased demand for freight delivery travel'. This was not a view shared by Sydney and Perth interviewees.

Finally, two statements explore the 'separate markets' storyline. Of these 'The design of [city] transport systems must recognise that separate markets exist for different modes of travel' (Q30) received a uniform mean score of agreed in all three cities. This is interesting in that it appears to reflect the path dependence in the thinking that emerges from different agencies serving different modes (essentially separate silos), and conflicts with the noticeable twenty-first-century shift in policy discourse towards 'integrated transport' (see Chapter 7). The latter idea, however, received support in the much lower level of agreement that these different markets must be served by different transport agencies (Q31).

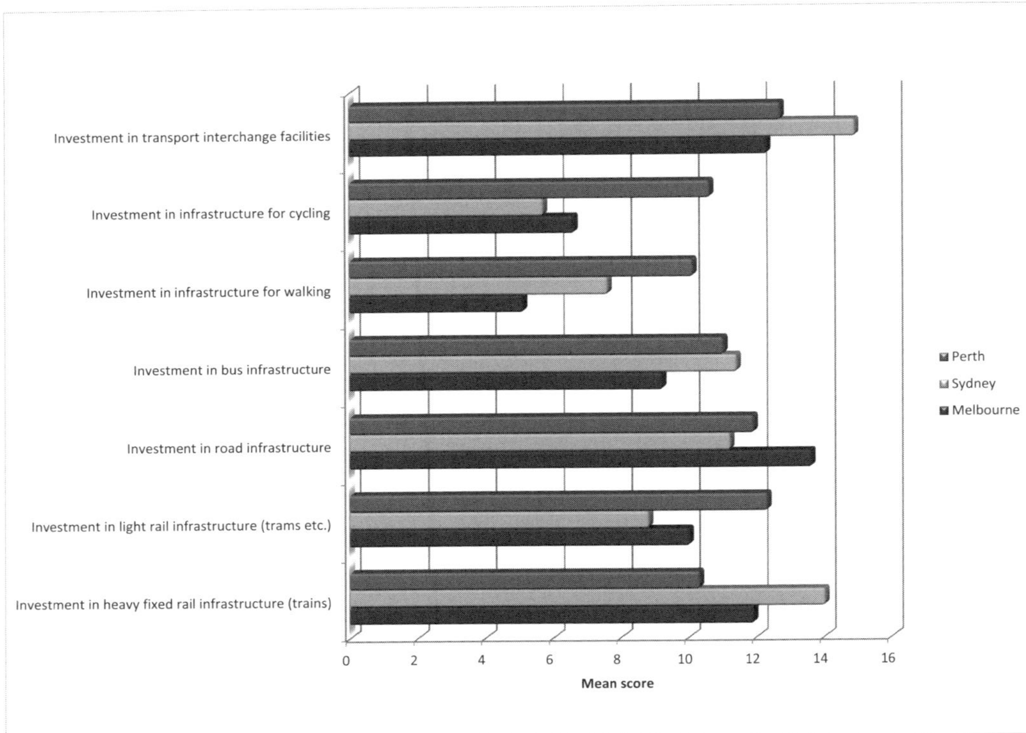

Figure 11.3 The contributions different forms of transport infrastructure make to growth in employment opportunities

The Town Planner's Tale

In Chapter 8 we noted that, in published policy, the town planning discourse has served to reinforce the Engineer's Tale and the Economist's Tale rather than providing the leadership needed to guide the form of the city and its supporting transport infrastructure. When examining the mental models of the policy shapers, this position continues to be the case in many instances (see Figure 11.4). That 'town planning should seek to reduce the time travelled between the home and other destinations within the urban system to which people choose to travel' (Q33) gained a similar mean level of agreement in all three cities. There was a lower level of agreement with the storyline that Australia's low density cities can only be served by private car (Q34 and Q52). This was particularly so for Sydney interviewees. The discourse is, however, more nuanced when drilling down to the detail. For Perth and Melbourne, there is agreement that 'investment in roads in outer areas is necessary to compensate for the lack of public transport opportunities in these areas (good public transport service being uneconomical to provide in such areas)' (Q35). This demonstrates that the roads policy pathway had not yet faltered. This is reinforced by a belief that 'building [motorways/ primary distributor roads] in outer urban areas where land reservations have been established is acceptable because it involves minimal social and environmental costs' (Q11). This was certainly the argument put in relation to the building of a section of Perth's orbital freeway (Chapter 8). This contrasts with the view that 'alternatives to the freeway are needed to provide better opportunities for cross-metropolitan travel' (Q37), which was not strongly agreed by policy shapers. There is also recognition that 'building new [motorways/primary distributor roads] in inner urban areas involves social and environmental costs that are too heavy for communities to bear' (Q10). This statement received a mean score of almost five (agreed slightly), indicating that there is not yet strong disagreement with this statement.

In relation to urban form and structure (Q38–Q40), there was a high level of agreement in all three cities that 'suburban development should be concentrated in spatial corridors around transport spines' (Q39). The differences between the city's policy shapers were more marked in relation to the other two urban form statements. Melbourne interviewees showed a strong agreement with the 'growth pole' storyline – 'In the outer suburbs better roads are needed to link households with locations of concentrated employment' (Q40). Sydney interviewees showed little agreement with the 'CBD vitality' storyline – 'The continuing vitality of [city] Central Business District depends on good road connectivity with the suburbs – bringing in customers to the CBD retailers and workers to CBD offices' (Q38). Sydney policy shapers appeared to identify strongly with their published transport and land use policies (see Chapter 8) in this respect.

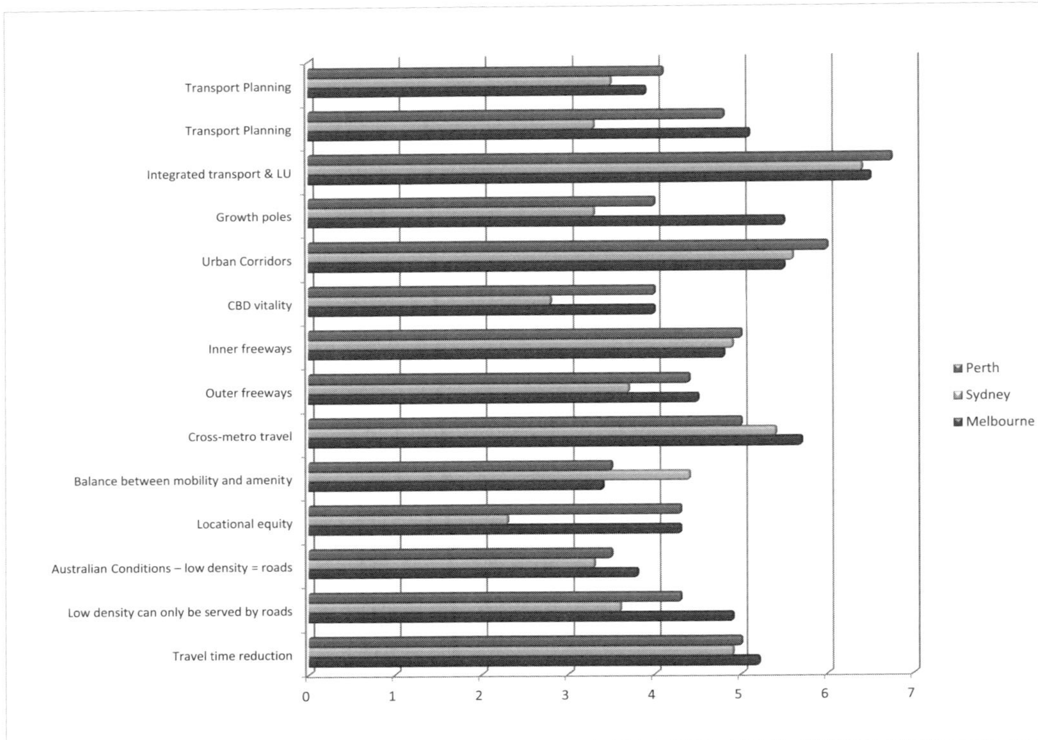

Figure 11.4 Mental models: The town planner's tale

There is a constant discussion among policy shapers and planners in Australia as to whether land use planning should lead in the shaping of the city, or whether transport infrastructure should shape the city. In our view a sustainable plan for city growth requires that land use decisions made in the context of environmental sustainability should shape the city, and transport infrastructure should be designed in order to support this plan, not the other way around. We aimed to explore this proposition with the statement – 'in planning a city, transport infrastructure should provide an integrated land use and transport system' (Q8). Policy shapers in all three cities indicated the highest level of agreement with this statement amongst all those in the Town Planner's Tale – further indication that 'integrated transport' is today a significant goal (see Chapter 7). Exploration of more concrete examples of land use transport integration, however, exposed a concern about the way policy shapers would tackle the matter. Two statements explored this concern (Q41, Q42). It is evident that for the policy shapers from Perth and Melbourne there was a higher level of agreement with the former statement – that 'the location of transport infrastructure investment should determine the location of development' – than with the latter (Q42). Policy shapers in all three cities were more closely aligned in their response to the latter statement – 'the need to ensure all new development is located where it is accessible by all modes of transport should take priority over other competing planning considerations' – but with a mean score of four this suggests a need for a deeper understanding of the real implications of land use transport integration if accessibility planning is to gain any traction.

The Sustainability Tale

A set of statements was used to explore the extent to which different elements of sustainable transport were accepted by policy shapers. What is interesting in relation to these items is the higher levels of agreement with many of the items in this Tale when compared to the other three (Figure 11.5); five of the sustainability items scored a mean of six (agree) or close to, with close consistency across all three cities.

The opening statement – 'an ecologically sustainable [city] cannot be achieved while governments continue to favour investment in road transport over public transport and other modes such as walking and cycling' – was agreed in all three cities, but the mean score was fairly low (at, or under, 'agreed slightly') demonstrating that the transport policy path has yet to be fully re-directed. If we tie this storyline back to the views above, it is evident that a roads based transport solution is still seen as appropriate for Australian cities, although this is now more moderated, with a view that, for example, the demand for car travel must be managed through road pricing.

On resource efficiency, two statements aimed to draw out this storyline (Q56, Q57). The first focussed on fuel scarcity as a rationale for cities to adapt their transport systems for non-car modes, and this achieved a mean score close to six (agreed) by Perth policy shapers, with Sydney a close second and Melbourne slightly lower. The second statement focussed on the way differences in the way travellers pay for their use of cars and public transport incentivises car use.

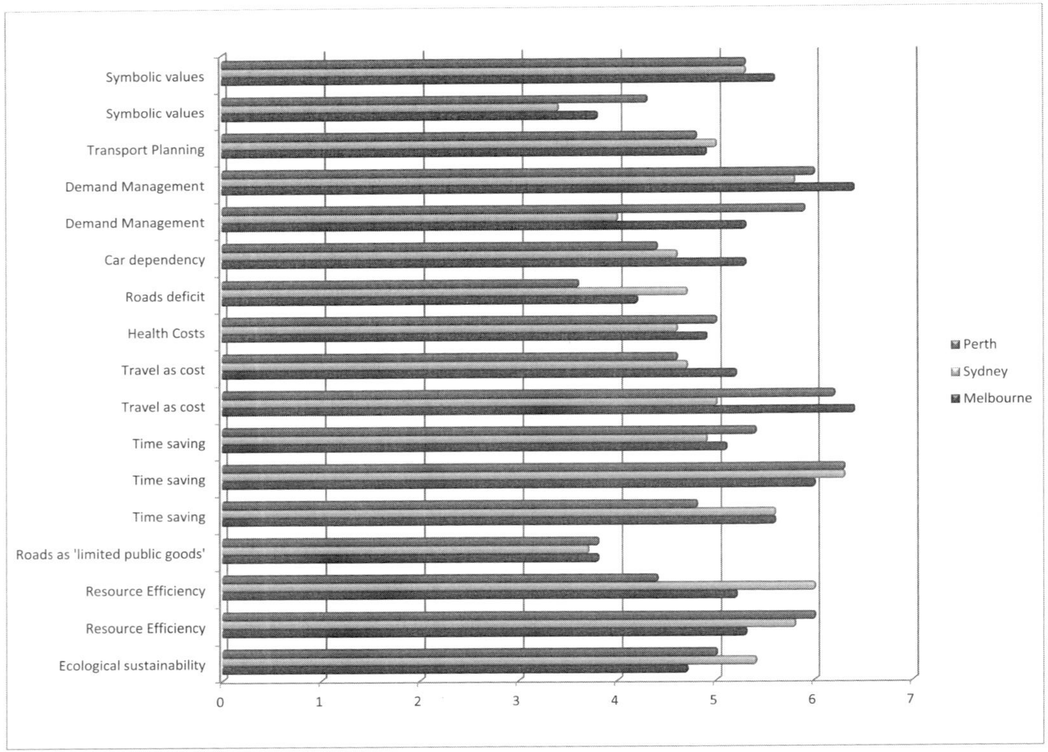

Figure 11.5 Mental models: The sustainability tale

Sydney's policy shapers gave this a higher level of agreement than Melbourne; Perth interviewees were closer to neutral.

The Economist's Tale can be modified in such a way that environmental sustainability can be factored in. We did this in several statements (Q58–Q62, Q66, 68, 69). The first of these statements asserted that freedom of movement for people and freight could not be achieved without costs (severe traffic congestion, loss of lifestyle and environmental values) – this was not accepted by our policy shapers where the mean score was between three (disagree) and four (neutral). That 'economic benefits could be achieved by saving time by reducing unnecessary travel' (Q59) was agreed by all three policy sharper groups, slightly less so in Perth. This suggests that use of travel time savings in cost-benefit analysis for transport infrastructure projects should be dispensed with, or at the least where they are used a qualification of which forms of travel are necessary should be made and only these accounted for. Gaining economic benefit through use of advanced logistical planning to reduce unnecessary freight travel (Q60) achieved one of the highest mean scores in the Sustainability Tale. Saving time by teleworking and teleshopping (Q61) was also slightly agreed (with a mean score of 5). There was also slight agreement that the current costs to life and public health from road transport are so high as to necessitate a change in investment priorities (Q62). There was agreement that 'The full costs of travel (in terms of lives lost and injuries from road accidents, health damaged from pollution as well as the isolation of those left without car access in a car-dependent city) should always be accounted for in any benefit cost analysis' (Q68), but less agreement with car users paying the full costs of car use (Q69). We concluded this exploration of the alternative Economist's Tale by examining the 'roads deficit' as an alternative to the 'public transport deficit' storyline (Q66), this did not receive any agreement, so while the public transport deficit is a storyline receiving less support than appears in the policy documents, there is no willingness to place public transport and road based transport on a level playing field.

Turning to town planning in a Sustainability Tale, the statement that 'the city has become increasingly 'car-dependent', meaning that the only viable transport mode choice for most people is use of the private car', was only slightly agreed with – more so in Perth, less so in Melbourne. Instead, Perth policy shapers showed a higher level of agreement that 'reducing the need to travel (on all modes) must be a key part of any transport or urban planning strategy' (Q64), as did Melbourne. Sydney policy shapers were neutral on this point. 'Managing demand for car travel – and seeking to reduce demand – should be a necessary part of any urban transport strategy today' (Q65) was one of the statements that received a high mean score (close to 6 or above), with Melbourne policy shapers scoring the highest mean. There was limited support for the statement that collectively the public should choose the city form and then a transport system be designed to create this (Q67).

We concluded the Sustainability Tale with two statements about the symbolic values of car use. There was little agreement that 'People choose cars over other modes of transport because of the symbolic values they embody (e.g. sex appeal, escape, exciting lifestyle, wealth, freedom) at least as much as for the convenience they provide' (Q70), although we doubt that those responsible for selling cars would agree!. There is greater agreement that when people do change their travel behaviour away from car use this leads to changed valuations of modes of transport. This suggests that providing city dwellers with the opportunity to use alternative transport modes will be an effective way of engendering support for these modes.

The Top Storylines

Six storylines achieved a mean score of 6 (agreed) or over (strongly agreed) and there were little differences between the collective views of policy shapers in each city (Table 11.1).

Table 11.1 Mental models: Top storylines

Tale	Storyline	City perspective
Engineer's Tale	Balanced investment in transport infrastructure by mode (Q7a)	All three cities
Economist's Tale	Infrastructure investment brings employment growth (Q26)	All three cities
Town Planner's Tale	Land Use Transport Integration (Q8)	All three cities
Sustainability Tale	Demand Management – reducing demand for car use (Q65)	Melbourne (Perth and Sydney close)
	Cost-Benefit Analysis should account for the full costs of travel (Q68)	Only Perth and Melbourne
	Economic Benefits to firms can be achieved by reducing unnecessary freight transport through advanced logistical planning (Q60)	Only Perth and Sydney (Melbourne close)

It is here we see that there may be 'light at the end of the tunnel', as these storylines suggest the potential for a transition to a new pathway. There appears to be a shift in the mental models, at least away from the core storylines which have determined the traditional logic found in published policy for so many decades of road building over all else. That half of the most agreed storylines form part of the Sustainability Tale suggests that a new paradigm of sustainability may be behind this shift. We saw this, also, reflected in the later policy documents published at the turn of last century. Of course, these six storylines, in reality, cannot be considered in isolation from the support for many of the long standing storylines which have seen the path continue. It is this very interplay between storylines that 'dim the light', making the possibility to deliver change in transport policy through implementation a difficult one. What will be required is strong leadership by policy shapers to break the path and ensure that the new paradigm of integrated transport succeeds in setting a distinctly new path rather than simply being subverted and co-opted in support of the old.

The Top Storylines by Functional Group

The results thus far report the mean score for the policy shaper group as a whole in each city. The rationale for taking this approach is that, in our experience, the views of each policy shaper have an influence in decisions made on transport planning. In this way it makes sense to view the result for the group as a whole. In addition, we also explore the extent to which views differed according to the particular 'functional group' that the policy shaper belonged to. For each city we grouped the policy shapers into one of five groups: A) agency bureaucrats – 1) roads; 2) public transport; 3) town planning; and B) office holders at the political level – 4) state politicians; 5) local government chief executive officers. With such a small number of individuals in any one functional group, generalizing from the results must be treated with caution. For this reason only the strongest level of agreement for each group is examined. What we show here is the results for the 'agreed' (6) or above (Figures 11.5–11.7).

The extent to which storylines from the Sustainability Tale feature varies between each city: Perth records the greatest number of these at fourteen, Sydney at ten and Melbourne at nine. When the 'activists' (bearing in mind that activists represented both pro-roads and anti-road/pro-public transport) are removed from Perth's result the number of sustainability storylines drops to ten. It would appear, then, that the Perth policy shapers were more predisposed towards the sustainability mental model. Of the other three Tales, in each city it was the Town Planner's Tale that was most strongly supported, with all groups drawing on this Tale (with the exception of Melbourne, where politicians did not score this Tale at six or above).

Comparing the support for storylines by bureaucrats from each city an interesting picture emerges. In Perth the roads-based interviewees are closely aligned with the storylines from the Engineer's Tale, whereas in Sydney this group draw on a much fuller range of the Tales, but they score the storylines in the Engineer's Tale the highest. This suggests a need to focus professional education on re-training for those in the roads-based agencies, but this would need to be extensively researched in future work. In Melbourne we see a role reversal, where roads-based interviewees rarely rate highly storylines from the Engineer's Tale. This may be the result of how the sample of interviewees was structured with the forced exclusion of VicRoads (see above). Town Planners draw on a range of storylines from all four Tales (Figures 11.6–11.8).

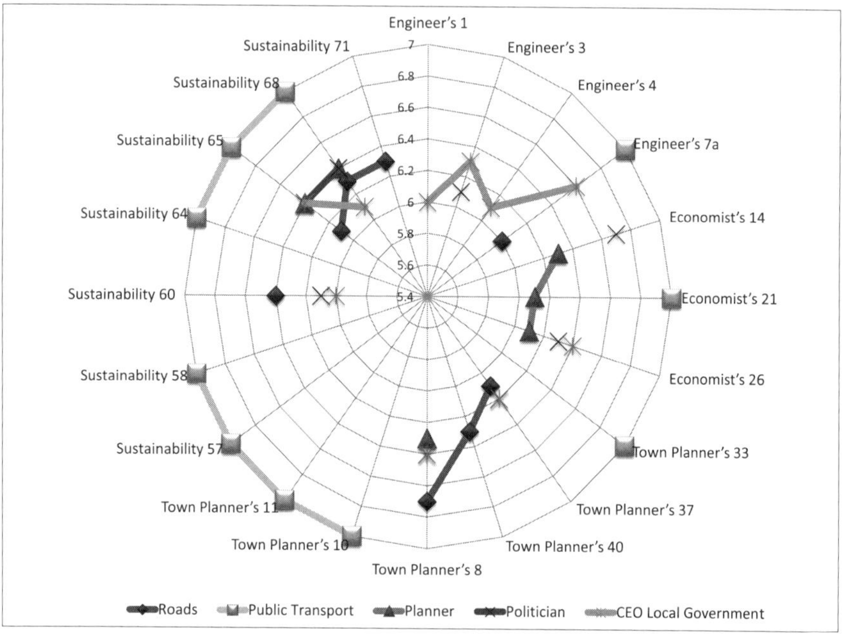

Figure 11.6 Top storylines by functional group: Melbourne

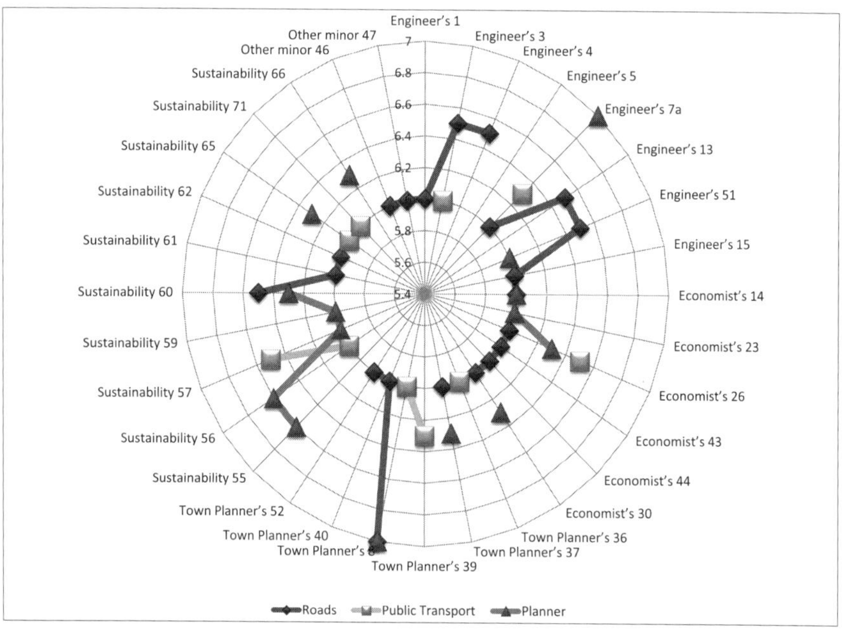

Figure 11.7 Top storylines by functional group: Sydney

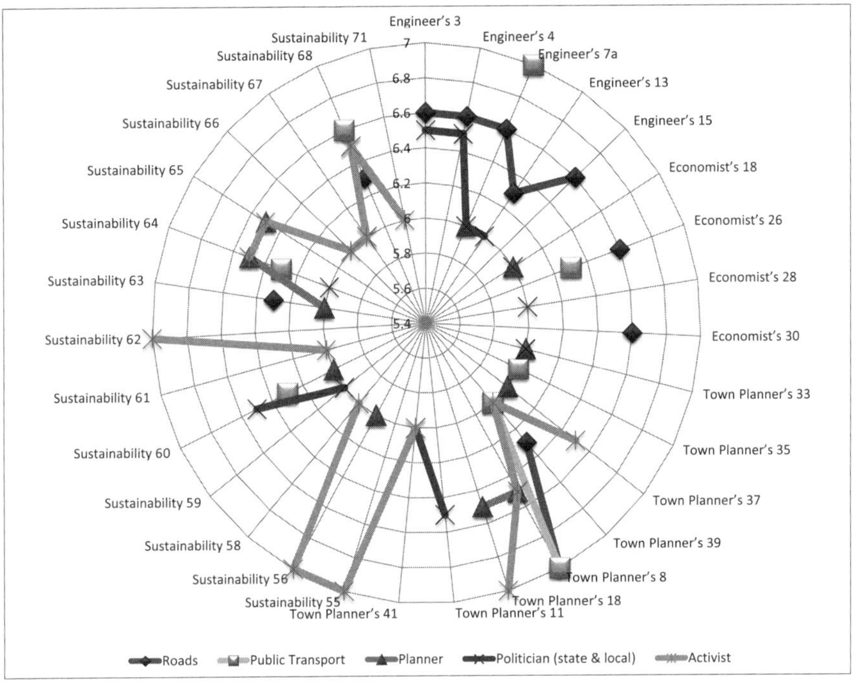

Figure 11.8 Top storylines by functional group: Perth

Comparing the top storylines of roads-based interviewees with public-transport based interviewees highlights an interesting pattern. In Perth and Sydney, roads-based interviewees draw on the full gamut of the four Tales (in Sydney this is pronounced) – whereas public-transport based interviewees draw on a much smaller number of storylines (see Figures 11.9 and 11.10).

Conclusion

The strongest conclusion is that policy documents are much more than mere window dressing. There is a strongly suggestive, if not yet statistically tested, correlation between what the policy documents say and how policy shapers think. Apart from political action that sometimes dislocates the policy path, it is reasonable to assume that how policy shapers think determines how they act, for instance what advice bureaucrats and consultants give to politicians, and what the political tier expects. On the whole, the core storylines of the old paradigm of transport planning still hold sway in the minds of the shapers. Yet it is also clear that cracks are appearing in the logical consistency of that discourse. There are continual efforts to reconcile the storylines of Sustainability and Integrated

Transport with the stories of the old paradigm: predict and provide, balanced transport, investment in roads brings economic growth, the public transport deficit and so forth. Attempts at reconciliation result in cognitive dissonance which becomes increasingly difficult to conceal.

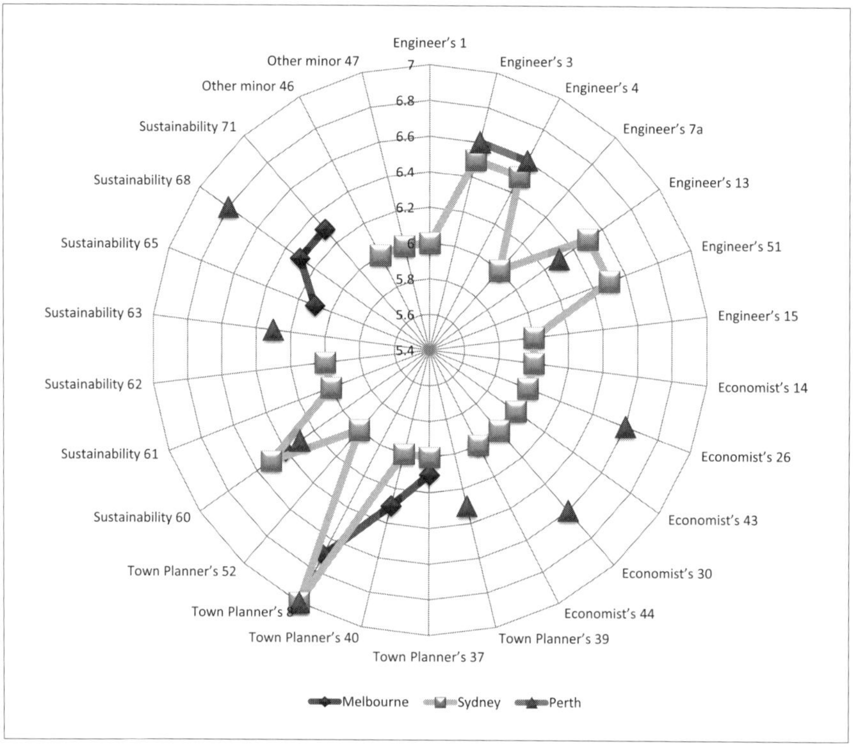

Figure 11.9 Top storylines by city for roads-based interviewees

Path dependence is therefore evident. While there is some acceptance that the roads solution needs to be modified by demand management, the shift is relatively weak and does not extend to seriously limiting car use. The Sustainability Tale is itself not strong or entirely consistent. It is far from clear exactly what 'sustainability' means other than a general desire to prioritize environmental quality above mobility. It is too easy to co-opt sustainability to old paradigm ends. Those who adopt 'sustainability' need to specify in measurable terms what sustainability means. Likewise operationalizing land use transport integration needs to be spelled out; otherwise the new rhetoric will be subverted to support the old.

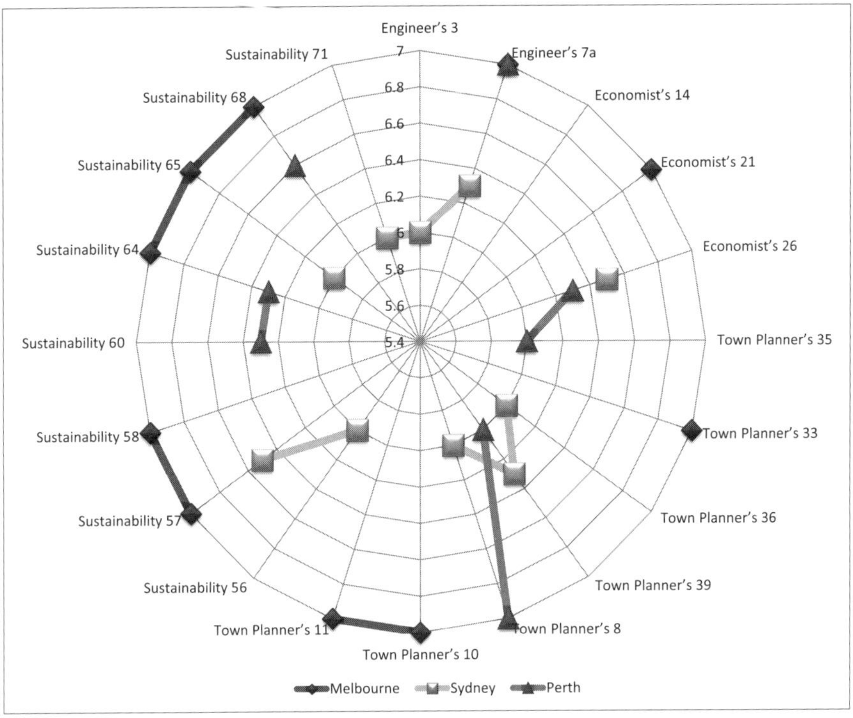

Figure 11.10 Top storylines by city for public transport-based interviewees

The observable difference between cities in response to the storylines is not great, which suggests that there is something of a national epistemic community here in which people share their views about professional issues. There is a small difference between Perth and the other two cities we surveyed. We postulate that the fact of substantial new investment in railways, and operationalizing of a form of integrated public transport (buses, bicycles and trains) has had an effect on the discourse. Also that may itself have resulted from community pressure allied to political intervention. So there is hope that a more widespread and far reaching critical juncture in the discourse that will dislocate the old path is beginning to unfold.

Chapter 12
Overcoming the Barriers

There appears to have been a wave of thinking about transport policy propagating throughout the developed world, and now reaching Australia's shores, that, wherever it strikes generates 'tipping points' of change. In Australia today the balance of public opinion, expert opinion and even political opinion is moving away from the ideal of unlimited car use for routine trips. This movement is being driven by the failure of massive and costly programs of construction of urban motorways to deliver congestion free motoring, the prospect of high population growth in the cities promising more traffic chaos, and the rising price of fuel stimulating an unprecedented rise in public transport use – with concomitant public dissatisfaction with public transport systems essentially unchanged since the beginning of the last century. Even utilitarian cycling (for the journey to work, school and shop), as against cycling for health and leisure, has had something of a renaissance. In the background is the always turbulent politics of climate change, with fossil fuelled transport in the spotlight, and the steady encroachment of ideas contrary to the conventional wisdom about Australia's transport future. But the conventional wisdom and its institutional matrix create barriers to change.

This final chapter, then, has two main tasks: to sum up what we have learned about institutional barriers, and to propose what might be done to change the policy path that has been so dominant over the last 60–70 years. First, then, we sum up what we have learned.

Institutional Barriers to Sustainable Transport

Sustainability is possible. It refers to a range of social, economic and environmental improvements. But will gradual change suffice? A transport system that contributes substantially to global warming and the consequent destruction of the biosphere as we know it is profoundly and absolutely unsustainable. That is what we have today. The theory of path dependence provides an explanation of why things stay the same in the face of the need to change. Path dependence applies to patterns of production and consumption, to the facts of urban built form, and to people's habitual behaviour. In transport, we also find path dependence in institutions: in this case the institutions surrounding transport planning. Institutional path dependence has an organizational and a discursive aspect – the strength and weakness of organizational structures, and the talk that explains, justifies and shapes public policy. If discourse changes other changes may follow.

As we discussed in Chapter 4, roads and cars are deeply embedded in the Australian economy as well as in Australian popular culture. Australian transport policy has too often been compartmentalized geographically, institutionally and functionally. While the delivery of metropolitan transport services and land use planning is focused at State level, much power to determine outcomes lies at federal level through its command over large scale funding. At federal level the politics is not focused on service delivery, or on urban development, but on the national economy and strategic hand-outs at budget and election time. Australia may have reached a new critical juncture. There are major new urban and regional rail projects and light rail projects being implemented across the nation. There are useful bus rapid transit projects and new ways of using buses to provide connections to local communities too long isolated. In recent times Infrastructure Australia and the Commonwealth Government have shown that new freeways do not have to be the centrepiece of urban transport planning and investment.

At State level there has been a historical flux of organizational consolidation and fragmentation, as we showed in Chapters 5 and 6. Generally the roads agencies have added to their well-established professional competence an increase in their scope of operations and organizational power. High levels of long term funding have enabled them to engage in continuous planning of new infrastructure. Sometimes their plans have been politically checked, but then the plans have been adapted to what is politically possible. Public transport organizations, originally arranged in functional silos, have seen periods of consolidation. But this movement has been turned back in Sydney and Melbourne by the neo-liberal policies of corporatization and privatization which again brought renewed operational fragmentation. How to plan transport in a system structured by neo-liberal dogma that denies the virtue of public planning remains a puzzle in these cities.

Consolidation of public transport operations and its engagement with infrastructure planning seemed to be promised in Melbourne in the early years of the Cain Government (from 1982). But the managerialist tenor of Cain's strategy meant that political action was so focused on structural reform of government that there seems to have been little room to think about reform of policy. To use a football metaphor, the Cain Government was focused on strategy to get hold of the ball but did not think so much about where to run with it. In Perth the politics of transport took a very different turn. The movement to reinstate railways was driven by a political-bureaucratic alliance that eventuated in the construction of major new railways, the establishment of central planning of routes and timetables, and the operational integration of transport modes.

In all three cities the roads agencies remain integrated, powerful and well-resourced units within the departmental structure. In contrast the public transport agencies have been deliberately fragmented. In Sydney and Melbourne service delivery is separated from strategic policy (corporatized or privatized). In Perth the state agency retains strong contractual control over service delivery, despite the fact that bus services are delivered by private companies. In Sydney there was certainly a political will to reinstate public transport as a priority but instead of resulting in firm and agreed action that could stay the course, the political will was internally conflicted and resulted in constant vacillation over rail infrastructure projects. As we explained in Chapter 7, a movement towards integrated transport planning is clearly apparent in all three cities to varying degrees, and with varying methods and emphases. Integration of modal planning is implemented in Perth. The aspiration to integrated transport planning is now enshrined in law in Victoria and New South Wales (and therefore Melbourne and Sydney). The integrative movement has a long way to go before real operational, system-wide integration, and especially integrated land use and transport planning, is achieved. But there is enough in what we have found to suggest that a critical juncture has been reached in which the path towards integration will be hard to reverse.

In Chapters 8 and 9 we described the structure of discourse in the roads and public transport sectors. For much of the twentieth century the dominant storyline around public transport was about how public transport, especially rail transport, was obsolescent. The 'public transport deficit' occurred because, from mid-century, people were leaving trains, trams and buses and using cars – choosing to shoulder the burden themselves of paying for private transport. This left public transport performing a residual function, one important aspect of which was reducing traffic congestion on the roads. But this effect was not much noticed. Instead the story that public transport performed a welfare function was used to justify the deficit. Meanwhile the roads agencies tackled congestion by building more and better roads, which of course attracted more people to use them. The storylines of obsolescence, deficit and welfare presented public transport in a negative light. Only bus transport to some degree escaped this negative cast because buses run on roads and have some of the flexibility of private cars. The roads policy discourse, and its associated tools and techniques, reinforced the implementation of a roads based transport system – one designed for high speed private car travel. There is a circular process in which planning and building new roads (institutional path dependence) shapes a more car-dependent and sprawling city (technical path dependence), and shaping of the city, literally set in concrete, reinforces institutional car-based solutions to the need for mobility and access. As we said in Chapter 8, planning for public transport is the Cinderella of transport planning. Planning of new rail infrastructure was for many years rendered nearly impossible by the pressure to reduce 'the deficit'. It is not so much that Cinderella was absent, but much more that she was denied equality in the transport discourse. Public transport was consigned to the scullery on the basis that she did not provide a worthy alternative to the car.

There is, however, sometimes a disconnection between what policy documents tell us as they unfold storylines and what happens 'on the ground', giving effect to political decisions. This can have positive results for sustainable transport, as in Perth when political leaders simply seemed to ignore the story that railways were obsolete. Sometimes the storylines exert a gravitational pull dragging back policy innovation to conform with the dominant path, as in Victoria when in the 1980s 'reducing the public transport deficit' overwhelmed the attempt to present the goal of an integrated public transport system in a positive light as being essential to a liveable city, and to work to implement such an outcome.

The supportive or contrarian power of stakeholders in relation to the conventional path was explored in Chapter 10. We demonstrated the strength of the pro-roads groups which operate as part of a 'discourse network', sharing a common argumentative language with their respective State government roads agencies. We examined the take up of the 'sustainability' discourse in the hands of the pro-public transport groups advocating a shift away from road building. If one looks to stories in the media as an indication of the relative power of these stakeholder groups one may be forgiven for thinking that a more or less equal balance of power exists between pro-roads and pro-public transport groups. Such contrarian groups are given much attention because they create controversy, which is the bread and butter of the press. Yet one only has to consider the size of membership of the automobile clubs, the immense financial resources they command, and the scale of their continuously operating media assets churning out the pro-roads storylines, to understand that there is a far from even balance of power. While their membership is held captive by the very real need for roadside assistance for car drivers – which is very well provided by the clubs, the stated main mission of these clubs is to advocate for the car and its users, a mission always carried out with zeal.

Of great interest, then, is the surprising adoption of a much wider mission by the NRMA (in Sydney and NSW) articulating the need for sustainable transport. This dislocation happened because in devising its *Clean Air 2000* campaign, the Association opened its thinking to a wider range of stakeholders including its main partner, the Australian Council of Social Services. Once again, however, we observed a contrarian initiative drawn back to the dominant path when this innovative campaign on the future of transport was quietly dropped.

Finally in Chapter 11 we reported an investigation into whether and how the storylines in the policy documents exist in the minds of policy shapers. There is a view that policy documents and government reports are of little importance because they do not give a true picture of the actual thinking of those who shape policy. Our research refutes this proposition. Of course as we also demonstrated, political action is not always determined by the storylines of the dominant path. Politicians and even senior bureaucrats can sometimes cast these stories aside, listen to other voices and act accordingly.

However, on the whole, the core storylines of the old paradigm of transport planning are still accepted by the policy shapers. The 'sustainability' tale is weakly and inconsistently articulated. It is unclear exactly what 'sustainability' in transport means. Therefore the sustainability tale is easily co-opted to support the stories of the conventional path. Yet it is increasingly difficult to reconcile the old logical structures with emerging events: how to account for the sudden growth in public transport use, even in poorly planned and maintained public transport systems, with the story of car dominance and public transport obsolescence; how to reconcile the evident unsustainability of the current transport systems with public concerns about climate change and rising fuel prices? How to reconcile promises of integration with actual organizational fragmentation? How to reconcile actual huge and continuous expenditure on motorways with investment in public transport merely promised at some future time? The public dissatisfaction with public transport in both Melbourne and Sydney played a significant part in the loss of government by the Australian Labor Party in both Victoria (2010) and NSW (2011) – the latter in a landslide. Reconciling the storylines of sustainable and integrated transport with the stories of the old paradigm – predict and provide, balanced transport, investment in roads brings economic growth, the public transport deficit and so forth – is becoming increasingly difficult. In the aftermath of privatization of public transport operations in Melbourne it is noticeable that the public transport deficit storyline has simply disappeared. This is not because there is any smaller subsidy from the budget to the operation of public transport, but because the neo-liberal remedy, privatization, simply did not produce that result. Similarly the fragmentation that resulted from corporatization of transport operations in Sydney has not resulted in a better service.

The Sustainability Tale is not strong or entirely consistent. It is far from clear exactly what 'sustainability' means other than a general desire to prioritize environmental quality above mobility. It is too easy to co-opt sustainability to old paradigm ends. Those who adopt 'sustainability' need to specify in measurable terms what sustainability means. Likewise operationalizing land use transport integration needs to be spelled out; otherwise the new rhetoric will be subverted to support the old.

So we now come to the question of what to do to overcome the institutional barriers to sustainable transport. We turn first to ideas about what causes transformative change.

The Nature and Conditions of Transformative Change

For planners and others interested in bringing about progressive change, the theory of path dependence raises a critical question: if public policy tends to run under its own momentum along lines determined by its history, how then can transformative change, *change of path*, occur? Plainly change, just as much as stasis, is enacted by agents: those individuals who populate and embody institutions, 'policy shapers' as we have defined them. As Torfing (2001: 289).observes, there is a 'false choice between the voluntarist idea that the world can be changed at will and the structural determinist view that institutional legacies determine the future'.

The theory of path dependence itself holds that new paths emerge at certain 'critical junctures' in which chance events play a definite role. But for planners the key question is whether new paths can be deliberately shaped. Nielsen et al. (1995: 6) speak of 'path shaping' arguing that 'social forces can intervene in current conjunctures and actively re-articulate them so that new trajectories become possible'. The idea of 'social forces', which can be a somewhat mystifying term, implies merely that change occurs as a result of individuals performing different social roles in pressing for change in circumstances where events outside the policy realm are throwing the policy path into question. One such role may be the organization of collective action – protests, demonstrations, pressure groups and so forth, but that is not the only or necessarily the most important role. The essays collected by Garud and Karnoe (2001a) elucidate a variety of such actor roles in 'path creation', for instance actors who embody new business interests, articulate new analytical technologies or create new organizational structures (Bassanini and Dosi, 2001), actors who construct 'transition paths' by creating and, managing 'niches' for new ideas within existing 'technological regimes'[1] (Kemp and Rip, 2001), entrepreneurs who 'exercise judgement as to how much of their deviations can be presented and communicated to key stakeholders in such a way that they are not threatened by them, but are instead galvanised by them' (Garud and Karnoe, 2001b).

Two political theorists, Hall (1993) and Torfing (2001), have explored the nature of transformative change, and have arrived at strikingly similar conclusions. Both are interested in social learning and institutional change, though they write from different theoretical perspectives. Hall (1993) conceives of policy change in terms of first, second and third order change. First order change is characterized as incremental change in the levels or settings of the basic instruments of policy. In the transport field first order change would be shifts in levels of expenditure by State and federal governments on different transport modes. The overall goals and instruments of transport policy would remain the same (for example fiscal subsidies for individualized transport, public investment in roads, budgetary expenditure on public transport services). Second order change occurs when the overall goals of policy remain unchanged but there are changes in the *means* used to attain them 'as a result of dissatisfaction with past experience' (ibid.: 278). In the transport field we might characterize the privatization of the management and delivery of road infrastructure and railway services as second order change, or a change in the instruments themselves (abolishing fiscal subsidies for individualized transport, or the introduction of congestion pricing). Third order (transformative) change is characterized by change in all three components of policy: the instrument *settings*, the instruments themselves (or *means*), and the hierarchy of *goals* behind policy. In transport, third order change might mean a change in which solving traffic congestion and the promotion of individual mobility are demoted in the order of goals and come to fall behind the goal of 'transport sustainability', read as

1 A 'technological regime' is defined as 'the grammar or rule set comprised in the complex of scientific knowledges, engineering practices, production process technologies, product characteristics, skills and procedures, and institutions and infrastructures that make up the totality of a technology (for example, an internal combustion engine or gas turbine) or a mode of organization (for example, the Fordist regime of mass production).' Within such a broad definition the production of an urban transport system could well be termed a technological regime.

addressing climate change mitigation and adaptation, and effective response to increasing scarcity of cheap fuel.

First and second order change, as we understand it, characterize shifts along the way but still within an existing policy path. Third order change refers to change in the policy path (paradigm or technological regime) itself. So how does third order change occur? Hall (ibid.: 280) makes three observations on the process of transition from one paradigm (or path) to another. First, the process of transition is unlikely to be made on the basis of expert views or rational arguments (that is on the basis of science) alone. The movement will ultimately entail political judgements. Secondly, the question of authority is central to the process of transition. 'Faced with the conflicting opinions of the experts, politicians will have to decide whom to regard as authoritative' (ibid.: 280.). And thirdly, as anomalies appear that cannot be comprehended within the existing paradigm, there will be *ad hoc* attempts to stretch the paradigm to cover them, 'but this gradually undermines the intellectual coherence and precision of the original paradigm'. As we have noted above the coherence of the transport planning paradigm is increasingly in question. So the transition is likely to be marked by policy experiments and failures 'that gradually undermine the authority of the existing paradigm and its advocates even further' (ibid.). Thus, Hall notes, first and second order changes do not necessarily lead to third order change. Rather, the movement from one paradigm to another 'is likely to involve the accumulation of anomalies, experimentation with new forms of policy, and policy failures that precipitate a shift in the locus of authority over policy' (Hall, 1993: 280).

Torfing (2001) also observes that transformative change is preceded by 'anomalies, puzzles and challenges to which the established policy paths have no adequate answers' (Torfing 2001: 288). He argues that a policy path is not merely a certain way in which some privileged decision-makers *choose* to regulate a given field of objects, processes and actions. Rather 'it forms a discursive terrain in which objects of regulation, the regulatory agencies, and the institutional forms of regulation mutually constitute each other' (Torfing 1998: 137–138). Thus 'the policy path is formed through a complex interaction between deliberate design, long lasting traditions, learning processes, and chance discoveries' (ibid.: 287). 'The policy path has the form of a narrative that leads us in different directions but continues to be organized around a few recurrent themes' (ibid.). When comprehensive institutional reform occurs, 'It is because the institutional framework has become destabilized and the institutional barriers to change have become weakened' (ibid.: 288). In some ways what we have described above displays a weakening in the consistency of the transport discourse, but in other ways the organizational structural bias remain strong.

Torfing argues that a policy path will have a certain elasticity that can account for and cope with exogenous events by 'mobilizing its discursive resources, stretching its interpretive schemes and modifying its rule governed practices' (ibid.: 288). But, 'There will always be a limit to its capacity to inscribe and domesticate new events emerging at the local, national or global scales' (ibid.: 288).

Dislocation of the policy path then occurs. This 'dislocation' of the policy path opens a window of opportunity for *political* action: 'The common experience of frustration, negation and loss will tend to construct a vaguely defined agenda to which the social and political forces must respond in order to speak in the name of society. The new political agenda will be structured around a series of empty, or at least floating, signifiers that aim to codify the absence of an adequate and legitimate way of unifying, organizing and regulating social, economic and political relations (Laclau, 1994)'. So we should not be surprised to find a new political agenda organized around the ill-defined (floating) signifier of 'sustainability'

Rather contrary to Hall, Torfing does not place so much weight on the existence of a strong alternative cognitive framework contributing to the change of path. And he also argues that a new path will display remnants of the old:

> The institutionalized ambiguity of the policy path, which stems from the fact that the path coexists with, and even articulates, residues from previous paths and embryonic elements of alternative paths, further sustains the possibility of institutional reform. In sum, 'institutional reform is facilitated by a complex interplay of internal sources of instability and external events that dislocate the structured coherence of the policy path'. (Torfing, 2001: 288)

A Program of Action

These concepts of transformative change throw light on the reasons why such change occurs and under what circumstances. Undoubtedly exogenous shocks, such as climate disasters like droughts, fires, rising seas, and floods, or sky rocketing fuel prices, will play an increasing role in policy change. But the appearance of such shocks is unpredictable and uncontrollable. Their impact may also be cumulative. Gilding (2011: 105) asks, 'What will be climate change's 9/11? What will trigger the shift? The answer is unsatisfying. While I am certain we will respond, it probably won't be triggered by a single event'. But in order to respond to the shocks with far sighted and beneficial policies there has to be human action, which is our concern here.

With acknowledgement of the path breaking work of Dudley and Richardson (2000), we can go a little further and identify six main components of action to promote transformative change that could be effective at a time when the existing policy path is under pressure or in process of dislocation thus providing windows of opportunity for political action.

1. The construction of an immanent critique of conventional wisdom.
2. A coherent and well-articulated alternative paradigm.
3. Politically effective individual actors.
4. The mobilization by these actors of networks of people with the capacity to exert influence.

5. The existence of (or creation of) authoritative institutions standing outside and above the existing bureaucratic apparatus: exogenous institutions.
6. Matching the timing of action to the dynamics of social change.

We conclude the discussion of each sphere of action with 'action guidelines', taking a leaf out of Rothman's book (1974) written at a time when social change was on the political agenda more than it has been ever since.

Immanent Critique

The dominant paradigm must be subjected to immanent critique – that is confronting the dominant path on its own terms, showing why the main premises of the paradigm are wrong. The most important and convincing arguments or 'storylines' (Hajer, 1995) supporting the dominant path must be identified and evidence adduced to show why they are wrong and can no longer be supported. An important Australian example of this critique is provided by the work of Mees (2000) who has demolished the argument that high quality, cost effective urban public transport could not be provided in low density residential contexts (as in the dispersed suburbs of Australian cities). The work of the Standing Committee on Trunk Road Assessment in the UK (SACTRA) provided evidence that building new roads created 'induced traffic' which, enhanced by the work of Downs (2004) showed why a massive road building program could not defeat traffic congestion, thus challenging the storyline that road building enhanced productivity and economic growth (other than from the economic stimulus that building itself created).

This kind of critique is different from much of the twentieth-century criticism of the roads programs from environmentalist and urbanist sources, (for example Newman and Kenworthy, 1989, 1999), important though the latter is, namely that building roads has negative environmental and social consequences. Without *immanent* critique, it can always be argued that the costs of road building can be traded off against the putative economic and social benefits. Immanent critique calls into question the long run benefits. On the cost side the environmental argument gains force when the future environmental costs can be shown to be overwhelming. The advance of climate science reinforces the argument for a radical shift to sustainable transport (see Moriarty and Honnery, 2010: 128–133). Similarly, the work of Aleklett et al. (2010: 1398) provides an immanent critique of the International Energy Agency's premise of continuing growth of oil production (see www.abc.net.au/catalyst/stories/3201781/htm accessed 6th May 2011).

An interesting example of immanent critique at local level occurred in the debate over public transport for Perth's southern suburbs. Before the decision to build the Perth – Mandurah heavy rail line (traversing a low density suburban corridor extending over 70 kilometres) a report was published, following an authoritative investigation, which advocated the use of buses to serve the suburbs (see Chapter 9). This proposition appeared consistent with the principle of a more 'compact' city and intensification of development in the urban penumbra of the central area and inner suburbs – superficially appealing to 'urban sustainability'. However this corridor was already served by a motorway and to compete with the car a high speed public transport mode would be needed. This could only be provided by heavy rail.

Understanding the multiple rationalities and causal forces at work in the dominant paradigm can contribute to immanent critique by exposing the institutional context of the dominant paradigm, including the use of path dependence analysis. As the dominant paradigm is faced by 'anomalies, puzzles and challenges' (Torfing, 2001), it will tend to generate alternative storylines in support, for instance now that it accepted that it is not possible to fix congestion for private cars by building more roads, the proposition is that roads must be built to facilitate freight movement (see chapter 8). Why then are 'freight only' or 'public transport only' roads rarely mooted and never implemented?

Action Guidelines

- Continue to identify the main storylines as they evolve which support the dominant paradigm. Subject these storylines to intense critical and objective scrutiny, bringing evidence to bear to challenge the main arguments in each storyline.
- Develop arguments for transformative change based on the future global environmental and resource costs of continuing the current direction.
- Make these arguments open and transparent, and demand transparency and testing of the specific arguments supportive of the dominant paradigm, for instance the assumptions currently hidden in the 'black boxes' of proprietary transport modelling. Demand open source models.

Construct and Articulate an Alternative Paradigm

Critique is not enough to dislocate an entrenched path of discourse. It is also necessary to construct and articulate an alternative path complete with new storylines supported by rational argument and as much evidence as can be adduced. For instance, the neo-liberal paradigm of governance was already constructed long before being used by political actors to transform economic management (for instance Hayek, 1944; Mises, 1963). It is useful to conceive of the alternative transport paradigm in terms of 'sustainability'. The very difficulty of attaching the word sustainability to a particular state of affairs means that the word is capable of drawing together a wide variety of actors and groups who share the awareness of a lack, something necessary which is currently missing. Laclau (1994: 174) states: 'This emptying of a particular signifier of its particular, differential signified is, as we saw, what makes possible the emergence of 'empty' signifiers as the signifiers of a lack, of an absent totality'. As we argue further below, third order change will require a very substantial coalition of actors and struggles over a wide front with potentially conflicting interests. Some way of unifying these struggles is almost certainly a requirement of success. This may seem Machiavellian but, as with 'sustainability', words like democracy and justice are just as hard to pin down, but also contain noble ideas whose real meaning is more likely to be determined in practice than in theory.

Therefore we use the term 'sustainable transport' (which is of course already widely used and familiar to many people concerned with transport planning) to describe a new transport paradigm that can stand up to the challenges of climate change and peak oil, and in contrast with the dominant twentieth-century transport paradigm.

The authors are partners of the GAMUT Research Centre in Australia. The research focus seeks to contrast 'sustainable transport' as a future paradigm in contrast with the dominant twentieth-century paradigm of 'auto-mobility' (see Cervero, 1996; Banister, 2005) and so change the way the community and governments view the future of urban transport. These terms are defined as follows:

Auto-mobility The twentieth-century vision of transport was one of private cars and trucks as the main way of getting people and goods around cities, and endless supplies of cheap oil. The big problem was 'traffic congestion', to which building bigger and 'better roads' was the solution.

Sustainable transport The twenty-first-century vision is one of seamless public transport networks even for dispersed urban areas, a reduced role for the private car for routine city trips, greatly improved facilities for walking and cycling, and integration of land use and transport planning to reduce travel distance and support non-car modes. In this vision, the main issue is seen as the impact of transport on climate change, increased oil prices and eventual reduction in oil supply. The growing public demand is for better urban public transport and its connection with foot and bicycle paths.

In the questionnaire administered to policy shapers we identified seventeen propositions of the new storyline of 'sustainable transport'. It should be noted that these propositions are intended for a 'developed world' context such as Australia, North America or Europe. They are as follows.

1. An ecologically sustainable city cannot be achieved while governments continue to favour investment in road transport over public transport and other modes such as walking and cycling.
2. Because fossil fuel is becoming scarce and more expensive, cities need to adapt their transport systems to make the best of alternatives to private car transport.
3. The way that car use and road space is paid for by the user compared with the way in which public transport is paid for (e.g. in fares collected at each journey) serves as an incentive to use roads.

4. Freedom of movement for all people and freight in cities cannot be achieved in the short to medium term without severe increases in traffic congestion, and loss of lifestyle and environmental values in the longer term.
5. Economic benefits can be achieved by saving time by reducing unnecessary travel.
6. Economic benefits to firms can be achieved by reducing unnecessary freight transport through advanced logistical planning.
7. Time can be saved for many people by working at home or shopping from home ('telework and e-commerce').
8. Because the current costs to life and public health from road transport are so high, a change of priorities for investment (i.e. from new roads to public transport infrastructure) is necessary.
9. The full costs of travel (in terms of lives lost and injuries from road accidents, health damaged from pollution as well as the isolation of those left without car access in a car-dependent city) should always be accounted for in any benefit cost analysis.
10. The full costs of car use including congestion, costs to life and health, pollution, and greenhouse emissions, should be paid by the car user.
11. There is a 'roads deficit': the costs to government arising from road use are at present significantly greater than the receipts to government in taxes and charges for road use.
12. Cities have become increasingly 'car-dependent', meaning that the only viable transport mode choice for most people is use of the private car.
13. Reducing the need to travel (on all modes) must be a key part of any transport or urban planning strategy.
14. Managing demand for car travel – and seeking to reduce demand – should be a necessary part of any urban transport strategy today.
15. The public should be allowed to choose collectively what sort of city they desire, and then a transport system should be planned that will create the kind of city desired.
16. People choose cars over other modes of transport because of the symbolic values they embody (e.g. sex appeal, escape, exciting lifestyle, wealth, freedom) at least as much as for the convenience they provide.
17. Changing travel behaviour through economic incentives and other means will lead to changed valuations of modes of transport [following the theory of cognitive dissonance in which it is argued that people generally tend to reduce inconsistency between their values and their behaviour. Thus changing behaviour leads to a change of values.

Even better than an abstract idea like 'sustainable transport' it is useful to find or create concrete examples of elements of this paradigm in practice. The creation of the bus rapid transport (BRT) system in Bogota, Colombia, was based on the pre-existing successful example in Curitiba. According to Prof. Juan Pablo Bocarejo of the Universidad de los Andes in Bogota, Bogota's mayor Enrique Penalosa was in close touch with Jaime Lerner who initiated the Curitiba BRT system (Bocarejo pers. comm.). Similarly the example of the Verkehrsverbund in Zürich has been used in Australia as a model of effective transport governance (Mees, 2000, Ch 5). These examples are geographically and even culturally distant from Australia. However the TransPerth public transport system, for example, owes much to the idea of network planning originating in Europe as well as an understanding of unsuccessful examples of privatization from the UK.

Action Guidelines

- Develop the idea of 'sustainable transport' in different social and geographical contexts creating a new paradigm of transport and mobility for the twenty-first century.
- Disseminate the new paradigm as widely as possible amongst political and professional actors, especially those in positions of power and influence.
- Find or create concrete examples of the new paradigm in practice.

Politically Effective Actors

The evidence from the cities around the world known for implementing a paradigm of transport different from the standard pattern of motorized personal mobility strongly suggests that leaders in positions of power play a key role, for example Jaime Lerner in Curitiba, Enrique Penalosa in Bogota, Michael Harcourt and Harry Lash in Vancouver, Alannah MacTiernan in Perth. Dudley and Richardson (2000: 229) write that,

> A key factor in understanding the paradox of the interrelationship between policy stability and change is to appreciate how new ideas may infiltrate and be absorbed by, or even bypass, interests and institutions in order to transplant themselves and bring about a paradigmatic shift in policy. In facilitating and executing the introduction of new ideas, it appears to be key individuals, acting in a variety of roles, who are primarily responsible for this task.

In the case of British transport policy these authors recount the part played by successive Ministers of Transport first in mobilizing support for the motorway program (John Boyd-Carpenter and Harold Watkinson), and subsequently in reshaping the policy agenda with a focus on the environmental problems and limits of the road building solution. Rail privatization in Britain, representing a similarly momentous policy shift, was also driven by politically powerful actors (Dudley and Richardson, 2000: 228).

Oppositional actors in the community play a critical role in transformative change. Michael Harcourt in Vancouver started out in the late 1960s as a lawyer acting for communities opposed to freeway construction. Harcourt was backed by a strong 'advocacy coalition' of academics and residents (The Electors' Action

Movement: TEAM), making his way on to the City Council and eventually becoming Mayor. Harcourt found a strong ally in Harry Lash, head of planning at the Greater Vancouver Development Authority. Dudley and Richardson (op. cit: 111) single out John Tyme in the UK as particularly effective, 'a zealous anti-motorway campaigner [who] made it his mission in life to undermine and eventually destroy the hegemonic roads policy which had held sway since the late 1950s'. Tyme was able to mobilize local NIMBY opposition around specific projects such as the Winchester By-pass to question the fundamental validity of existing transport policy. Tyme's strategy was to disrupt public inquiries into highway proposals:

> Tyme understood instinctively that the public inquiry was his arena and he would be weakened if he left it. Consequently, he displayed no interest in conventional direct lobbying of national government. Instead, coverage of inquiry disruptions via the mass media conveyed a clear message to policy makers. (Dudley and Richardson, 1998: 736–7)

Australia has had its own oppositional actors: Peter Newman in Perth and Paul Mees in Melbourne. Newman in his analysis of other international cities work with Jeff Kenworthy (1989, 1999) paved the way for the Perth's urban passenger railways. Mees, was president of the PTUA in Melbourne in the 1990s (see Chapter 10). He showed how networked public transport could provide high quality service to dispersed cities (2000, 2009). But whereas Newman found allies in the State Government of Western Australia: for instance Julian Grill at political level and Stuart Hicks in the bureaucracy, Mees was supported by community groups such as ARRO – Anti Ring Road Organization – in Melbourne's eastern suburbs opposing extension of a ring road across the Yarra Valley (Stone, 2008). His strategy, like that of Tyme was to use public inquiries to introduce an alternative agenda.

Of course if one were to rely on the already existing presence of committed politicians in positions of power (mayors or ministers) to bring about third order change, there would be few examples of change. But powerful figures can sometimes be persuaded to adopt elements of the new paradigm, such as Bill Rodgers in the UK or Julian Grill in Perth. An example of persuasion occurred in Melbourne in 2009. Swanston Street is an important civic axis running through Melbourne's CBD which links two universities, the State Library and the Town Hall. The street has been partially closed to car traffic since 1997 but is used by taxis, buses and delivery vans at all times of day and is open to cars after 6.00 pm. It is also a key route for trams and cyclists through the city. In 2007 a former leader of the State Liberal Party, Robert Doyle, stood for election as Lord Mayor of Melbourne. His platform included re-opening Swanston Street to car traffic. Once elected, however, he was persuaded to change his mind and became a strong advocate for the complete remodelling of the street, banning cars and creating a safe thoroughfare exclusively for public transport, pedestrians and cyclists (Dowling, 2011). Implementation of the plan commenced in May 2011.

Action guidelines

- 'Change Agents' in the community should either build broadly based advocacy coalitions to bring irresistible pressure on governments – sometimes becoming elected to power themselves, or forge alliances with government actors, helping them to construct their own political identities around the alternative agenda.
- It follows that change agents should be acutely aware of the political and social context of the time, including the opportunities this context affords for the implementation of strategies.

Actor Networks and Communities

Change Agents need the capacity to construct arguments in such a way as to facilitate coalitions, and agreement on action inclusive of a wide range of interests. Rothman (1974: 438) states that, 'the rate of adoption of an innovation is related to the degree to which it is perceived to be compatible with the existing values of the target system' [the latter being the person or group which the change agent wishes to engage in innovation]. Following Sabatier (1999), Dudley and Richardson (2000) discuss several occasions on which 'advocacy coalitions' were built around a change agenda. The environmentalists in the 1970s and '80s in the UK, these authors say, 'could build an effective advocacy coalition which could develop into a formidable adversarial policy community to oppose the roads community' (ibid.: 142).

In examples from Australia, the opposition to inner urban motorways in the early 1970s was forged through an alliance between new middle class residents of gentrifying inner city neighbourhoods, older working class residents, environmentalists and local councils. In Melbourne the united opposition of inner city municipal councils, including the City of Melbourne, was able to demonstrate the legitimacy of the protest to the more powerful State government. In Sydney a similar alliance was, unusually able to include even the trades union representing the workforce which would build the motorways, the Builders' Labourers' Federation (see Burgmann and Burgmann, 1998). In both cases the federal government was sympathetic to the opposition forces. In Melbourne the State government under Rupert Hamer capitulated. In Sydney, the State government was replaced at an election and the new government banned inner urban motorways. An advocacy coalition for sustainable transport was built by a most unlikely source: the NRMA, even though it was short lived (see Chapter 10). Unfortunately, contrarian forces in Sydney and Melbourne were unable to coalesce sufficiently around an alternative program to repeat their successes of the 1970s.

Action guidelines

- Change agents must forge powerful and extensive actor networks to develop and support an alternative to the dominant paradigm.

- Change agents must interpret the alternative paradigm in terms understood by and compatible with the interests of key groups, for instance, actors within bureaucratic agencies, local councils, local residents and business/professional groups.
- Change agents should seek ways of establishing the legitimacy of opposition coalitions, through broad representation of interests, and/or through (elected) local council support.

Exogenous Institutions

Dudley and Richardson (1998) make the point that established bureaucratic agencies tend to be resistant to innovation, a point already noted by Rothman (1974: 462): 'Organizations with many specific, strictly enforced rules are less innovative than organizations which have few rules and allow discretion in employee functioning' (a proposition supported strongly by six references in Rothman's survey of innovation). Dudley and Richardson (2002: 123), from their observation of British transport policy making, note that 'arenas without rules' – strictly not lacking rules but *outside* the normal rules – in the form of public inquiries played an important part in paradigm change. Commissions of inquiry, whether appointed by governments or initiated privately are able both to consider new definitions of transport problems and address evidence from new sources.

There are examples in Australia of similar 'arenas without rules'. Low and Power (1984: 182–184) describe two significant committees or boards of inquiry in Victoria set up by the Hamer administration in the 1970s: the Syme-Townsend Committee on administration of health services, and the Darvall Board of Inquiry into Metropolitan Government. Another important example was the Public Bodies Review Committee set up by the Victorian Liberal Government whose deliberations were fed into the platform of wholesale reform of the structure of government enacted by the succeeding Labor administration of John Cain junior. Since the 1980s, however, the approach of pluralist planning on the part of governments, opening up policy to new perspectives (termed 'values planning' by Low and Power, 1984) has been succeeded by an approach to policy making *within* the rules (at least in Sydney and Melbourne) increasingly subject to political direction and bureaucratic control. There have however been some exceptions including the Victorian Upper House inquiry into Melbourne's public transport instigated by the Greens, and the independent inquiry into Sydney's public transport auspiced by the *Sydney Morning Herald*.

Action guidelines

- Change Agents should seek out institutions outside the normal bureaucratic structures to advance their case for the new paradigm.
- Where possible Change Agents in positions of influence should help to create such institutions, such as public inquiries, parliamentary commissions and community forums.

Timing and the Dynamics of Change

Time is a critical factor in social change, and central to the thesis of path dependence. (Pierson, 2000b, 2004; Dudley and Richardson 2000: 239). The theory suggests that the process of third order change takes a long time to unfold. Even though visible change may occur quite suddenly, there is always a long period in which alternatives are prepared and postulated, opposition to the dominant paradigm mounts, and the logic of the paradigm begins to crumble. One or more external events may be the trigger for change but the trigger will not be pulled unless there has been a preceding confluence of pressures from advocacy coalitions advancing new ideas.

Anecdotal evidence suggests that the way in which the Transmilenio bus rapid transit (BRT) system in Bogota, Colombia was decided upon provides a good example of the good timing of Mayor Penalosa. The time for strong action was ripe. The previous Mayor Antanas Mockus had created a climate of public opinion accepting of change (and a successful example of innovative social policy). Political pressure to build an effective rapid transit system was strong, but agreement amongst key stakeholders on an expensive metro system could not be achieved. At this critical moment Mayor Penalosa was able to garner international finance behind a BRT system which had already been tried out successfully in Curitiba, Brazil.

Action guidelines

- Change Agents should not expect a smooth and continuous process of transition to a new paradigm but should maintain pressure over long periods in which the dominant paradigm is stable.
- Change Agents should pay particular attention to signals that a critical juncture is near in which opportunities for a radical shift may present themselves.

New Ways of 'Doing Planning'

This final section is addressed to the urban planning profession, particularly in Australia. According to the United Nations Environment Program (2011), cities worldwide produce 80% of the world's economic output, consume 75% of the world's energy and emit 71% of energy-related global emissions. While we have identified action guidelines that address the slow progress in third order change (above), this will not be enough. There is then an urgent need for innovative, path-dislocating or path shaping ways of doing planning. Planning must depart from the top-down, exclusive, centrally controlled, bureaucratic, routinized sort of planning seen in most Australian and many cities around the world (see March, forthcoming). In our study of policy discourse we show how planners have served to enact the ideas of road planners so reinforcing a particular car-based urban structure (Chapter 8).

The profession must become an advocate for real planning, aware of the vast challenges faced by future cities, challenges of adaptation to and mitigation of climate change. Urban planning, if it is anything, is about discovering and planning for the future needs of city people. It is not about simply projecting the past forward into the future, leaving almost everything to the land market and the development industry. Too often in Australia metropolitan plans look like bundles of infrastructure projects prepared by State governments tacked on to a range of 'hope' statements inserted amongst lavish displays of coloured photos. The public want truth and integrity in plans, and to see those plans realized accurately at local level.

To do this planner's must ensure they are widely informed, drawing on a full gamut of international research and planning practice in developing local planning policy. Planners need to enact the well-known principle: 'think globally, act locally'. There are now many examples and guides to low carbon policy-making from around the world, and these should become compulsory reading for policy shapers. Transport planners need to understand how renewable energy may be used for transport. For instance reports from governments and think-tanks around the world are showing that a zero carbon future is possible, incidentally demolishing a potent storyline that we have not considered in this book: 'green energy cannot be used for baseline power' and its corollary, 'only fossil fuel or nuclear energy can be used to provide baseline power' (see German Advisory Council on Global Environmental Change, 2005); Kemp (2010); Government of Denmark, 2011.

Planners must also be open to utilizing new analytical tools and approaches. For too long planning has been dominated by 'predict and provide' (Chapter 8) and has relied on the use of traditional four-step transport models to forecast demand for car travel based on past demand. Use of Cost-Benefit-Analysis tools has focussed narrowly on monetary costs, despite their ability to accommodate non-monetary values. Visioning and backcasting (see Banister, 2005) provide a different possibility, enabling communities to determine their vision for the future and plan a city backwards from this future end-point. Accessibility tools and models provide an output additional to traditional transport models for city planning based on accessibility by public transport, cycling and walking (see Curtis and Scheurer, 2010).

Planning practice must be open to a wide range of stakeholders. To do this the preparation and adoption of planning strategies and planning projects must be developed through open deliberation and be 'open book'. Planners must develop skills in facilitation and mediation (as well as the more traditional skills of analysis and design) in order to plan collaboratively with stakeholders. In Australia there have been recent examples of such practice. In Perth the production of the Network City metropolitan planning strategy (WAPC, 2004) followed a year- long process commencing with a one day 'Dialogue with the City' (Curtis, 2008; Legacy, 2010) where the community and a wide range of public agencies participated. In the end the very wide agenda for change in planning processes generated by those involved (see WAPC, 2004) proved too great a challenge for the bureaucracy who did not formally adopt the strategy and instead reverted to a centrally controlled and produced strategy – Directions 2031 (WAPC, 2009). Notwithstanding this procedural challenge, it is evident that a transition is occurring in the way in which land use transport integration is now conceived. Again in Perth, the introduction of 'Place-Based Planning' has been a means of breaking down the traditional working practices within silos of government, instead providing a process for government agencies to work together to bring a planning project to implementation (see Curtis and Punter, 2004).

The planning profession needs to regain the campaigning zeal it once had and encourage grass roots change. Political scientist Peter Self, in 2000 pointed out that, 'Both the state and the market are increasingly seen as the sources of remote decisions which bear harshly on the lives of ordinary people' (Self, 2000: 228). Planning worldwide needs to accept the challenge laid down by the 'transition town' movement, and many other local initiatives, whose stated aim is to equip communities for the dual challenges of climate change and peak oil.

Action guidelines

- Reform policy-making – plan for the future drawing on a wide range of international research and planning practice.
- Create and use new analytical tools for planning.
- Open planning to a wide range of stakeholders and plan collaboratively with them.
- Encourage grass roots change.

Our general conclusion from the above discussion is that third order change is a complex process, and change agents need to assemble a variety of strategies spanning long periods, and adapted to different conditions in different periods. These strategies need to be played out as an integrated whole. There is obviously much more that could be said about change strategies. What we have attempted to do here is to shift the attention of scholars and professionals inside and outside government to this process and the strategies that are arguably appropriate.

Conclusion

Urban planners *are* agents of change in all those professional domains that have an impact on cities (including engineers, economists, geographers, public health specialists, sociologists as well as land use planners). The history of the planning profession is one of dissatisfaction with the *status quo* and a commitment to change it for the better. It is in fact remarkable that a profession focused on cities and their development has been able to form around an agenda of change. In the academy, planning is also remarkable as being about more than description and explanation, but also about action. But the profession, like all professions has its conservative side, and, though there are exceptions (e.g. Sandercock, 1998) planning theory has only occasionally and rarely addressed the question of 'what is to be done'.

What has been described in this chapter and in this book may be disquieting for beliefs about what planning can be, namely the deliberate design of policy for the improvement of cities and society based on goals decided by the elected political tier. Much academic analysis in the last fifty years has demonstrated the limits of instrumental rationality and the interference in rational processes by social and political forces (e.g Lindblom, 1959; Wildavsky, 1973; Simmie, 1974). Yet, in practice, surely, we can say that the public still expect rationality and planners still try to deliver it. Indeed to abandon that expectation would be an atavistic reversal of the whole project of modernity and enlightenment.

However, in the problems confronted by mobility and transport today we are not in need of instrumental rationality flowing from pre-decided goals (Weber's *Zweckrationalität*, see Low, 1991: 75). Planning which follows a path through choice of policy to implementation may be more or less comprehensive, strategic or incremental, but change of the whole path and its direction follows an entirely different process involving everything that is political in the broadest sense in society. What we have tried to address in this book is the rationality of the goals themselves (Weber's *Wertrationalität*), the institutional barriers to change, and how basic change nevertheless can and does take place.

Can such change be deliberate? Certainly in the late 1960s and early 1970s some writers thought that social change could be 'planned and organised' (Rothman, 1974; Bennis, Benne and Chin, 1970). Today we have to return to that perspective in which planners join the ranks of 'change agents'. To overcome the barriers means that change agents must expect to take roles in a process with the following characteristics. First, change can take a long time from the launching of immanent critique, through the invention of an alternative paradigm, the discovery and emergence of trials and examples of the new paradigm, to its general adoption and widespread implementation. This is essentially a process of innovation with 'early and late adopters' (Rothman, 1974: 418–420 based on the work of Rogers, 1962). The time scale is very different from that of instrumental planning. Change agents must expect to participate in a process whose conclusion they may not see in their working lifetime.

Secondly, much action for change may take place which seems to be getting nowhere, or even going backwards, over long periods. Defeats are commonplace. However change occurs at particular times, 'critical junctures' in path dependence theory, when the time is right because a number of paths of action trodden by powerful actors converge. Change of path is therefore discontinuous, and can seem sudden and unexpected if the long processes which forge the converging paths are not seen. Often the arenas in which the new paradigm is expressed are outside the normal rules. Effective change agents take advantage of such arenas and may even play a part in creating them.

Finally, social change is always a collective process, not something that can be brought about by a single individual. Certainly there will be individuals who become recognized as champions of change, leaders of the new paradigm. But they do not stand alone. Change requires actor networks which change agents help to create and in which they participate, performing a variety of roles. Actor networks will include not only social activists and community leaders but also bureaucrats, politicians, professionals and academics. Building the new paradigm means building a supportive network which must eventually become dominant. There is a need for signifiers, such as 'sustainability', with which the new network is given meaning. Failure to recognize the connected nature of action, and the pursuit of individual glory, may result in isolation of the change agent and the generation of fear and increased resistance on the part of the actor network supporting the dominant paradigm.

The challenges all societies face today are economic, social and, at base, environmental. We are now up against the limits of the 'carbon civilisation' (Moriarty and Honnery, 2010). Transport and mobility is just one of many domains in which fundamental change will either occur deliberately or be forced on us. If it is forced on us the global, social and political results are highly unpredictable. But if there is to be deliberate action, then the barriers to change have to be understood and strategies for overcoming them pursued with the greatest urgency. Despite the exposition of long term change above, today we need to press the 'fast forward' button – we cannot afford to wait.

Bibliography

Adams, J.L. 1974. *Conceptual Blockbusting*, San Francisco: W.H. Freeman.

Aglietta, M. 1979. *A Theory of Capitalist Regulation: The U.S. Experience*, London: New Left Books.

Aleklett, K., Hook M., Jakobsson, K., Lardelli, M., Snowden, S. and Soderbergh, B. 2010. The Peak of the Oil Age – analysing the world oil production reference scenario in World Energy Outlook 2008. *Energy Policy* 38(3), 1398–1414.

Anderson, W.K. 1994. *Roads for the People, A history of Victoria's roads*, Melbourne: Vic Roads, Hyland House.

APT, 2004. *Action for Public Transport N.S.W.* [Online]. Available at: http://www.cs.usyd.edu.au/~jimd/apt/petrolheads.html. [Accessed: 25 February 2004].

Arendt, H. 1955. On Humanity in Dark Times, Thoughts about Lessing, in H. Arendt, *Men in Dark Times*, New York and London: Harcourt Brace, 3–31.

Arthur, W.B. 1988. Urban systems and historical path dependence, in Ausubel J.H., and Herman R. eds. *Cities and Their Vital Systems, Infrastructure, Past, Present and Future*, National Academy Press, Washington DC.

Arthur, W.B. 1994. *Increasing Returns and Path Dependence in the Economy*, Ann Arbor, USA: University of Michigan Press.

Ashton, N. 2003. The 1940s, The Beginnings of Planning for the Sydney Metropolis, in Toon, J. and Falk, J. 2003. *Planning or Politics, Town planning for the Sydney region since 1945*, Sydney: Planning Research Centre, University of Sydney, 2–48.

Ashton, P. 1993. *The Accidental City: Planning Sydney since 1788*, Sydney: Hale and Iremonger.

Astle, R. 2007. *How did we get here? An institutional analysis of the urban passenger transport sector in Melbourne: 1956 – 2006*, GAMUT Working Paper. Available at: http://www.abp.unimelb.edu.au/gamut/.

Australian Government Commonwealth Grants Commission 2006. *Issues Paper CGC2006/03, Architecture of Horizontal Fiscal Equalisation: Principles and Interpretations*, Canberra: AGCGC, March.

Australian Labor Party WA Branch. 1989. Australian Labor Party WA Branch. State platform 1989. Australian Labor Party, Perth.

Baer, P., Harte, J., Haya, B., Herzog, A.V., Holdren, J., Hultman, N.E., Kammen, D.M., Norgaard, R.B. and Raymond, L. 2000. Equity and Greenhouse Gas Responsibility. *Science*, 289(5488), 2287.

Ball, S. 1993. 'What is policy? Texts, trajectories and toolboxes' Discourse: Studies in the Cultural Politics of Education 132, 10–17.

Ball, D. 1996. *The Road to Nowhere? Urban freeway planning in Sydney to 1977 and in the present day*, Urban Research Program, Working Paper 51, Canberra: Urban Research Program, Australian National University.

Banister, D. 2005. *Unsustainable Transport, City transport in the new century*, London and New York: Routledge.

Bassanini, A.P. and Dosi, G. 2001. When and How Chance and Human Will Can Twist the Arms of Clio: An essay on path-dependence in a world of irreversibilities, in *Path Dependence and Creation*, edited by Garud, R. and Karnoe, Mahwah, New Jersey: Lawrence Erlbaum, 41–68.

Beale, Bob 1996a. Pollution message sinking in, *Sydney Morning Herald*, 9 May 1996, 11.

Beale, Bob 1996b. Smog wary commuters switching to bus, train, *Sydney Morning Herald*, 16 September 1996, 5.

Beder, S. 1997. *Global Spin, The corporate assault on environmentalism*, Melbourne: Scribe Books.

Bennis, W.G., Benne, K.D. and Chin, R. 1970. *The Planning of Change*, London: Holt, Rinehart and Winston second edition.

Bentley, I. et al. 1985. Responsive Environments: A Manual for Designers. Butterworth, Oxford.

Bettison, G. 1992a. South West Area Transit Briefing Note #1: An Overview. Perth: Transperth.

Bettison, G. 1992b. South West Area Transit Overview and Guide to Technical Information Pack II. Perth: Transperth.

Bettison, G. 1992c. South West Area Transit Project Director's Overview and Guide to the Technical Information Pack. Perth: Transperth.

Bevir M. and Rhodes, R.A.W. 1999. Studying British government: reconstructing the research agenda. *British Journal of Politics and International Relations*, 1(2), 215–239.

BITRE Australian Bureau of Infrastructure, Transport and Regional Economics. 2008. Australian Statistics Yearbook 2007, Table 4.3a, Total Passenger Kilometres Travelled per mode: Melbourne., Canberra: BITRE.

Black, W. 2001. 'An unpopular essay on transportation', *Journal of Transport Geography*, 9.

Blackshaw, P.W. 1974. *Sydney Area Transportation Study*, Sydney: The Australian Quarterly. 46(4), 56–68.

Blainey, G. 1968. *The Tyranny of Distance: how distance shaped Australia's history*, Melbourne and London: Macmillan.

Bocarejo (pers.comm.) This was a discussion Low had with Dr Juan Pablo Bocarejo, Director Grupo de Estudios en Sostenibilidad Urbana y Regional – SUR Coordinador general Centro de Investigación Estratégica en Movilidad Sostenible Universidad de Los Andes, on 11th May 2011.

Bottles, S.L. 1987. *Los Angeles and the Automobile, The making of the modern city*, Berkeley CAL: University of California Press.

Bratzel, S. 1999. Conditions of success in sustainable urban transport policy: policy change in "relatively successful" European cities. *Transport Reviews*, 19(2), 177–190.

Broomham, R. 1996. *On the Road, the NRMAs First Seventy Five Years*, Sydney: Allen and Unwin.

Broomham, R. 2001. *Vital Connections: A history of NSW roads from 1788*, Sydney: Hale and Iremonger in association with the Roads and Traffic Authority, NSW.

Brown, A. 1993. *Public Transport Reform – Victoria: From a System to a Service*; Government of Victoria Media Release, Melbourne.

Buchanan, C. et al. 1963. *Traffic in Towns, A study of the long term problems of traffic in urban areas*, London: HMSO the Buchanan Report.

Buchanan, J.M. and Tullock, G. 1962. *The Calculus of Consent, Logical Foundation of Constitutional Democracy*, Ann Arbor MI: Michigan University Press.

Building Better Cities 1993. *Newsletter Number 2*, August, Department of Planning, New South Wales.

Burgmann, M. and Burgmann V. 1998. *Green Bans, Red Union, Environmental activism and the New South Wales Builders' Labourers Federation*, Sydney: UNSW Press.

Campbell, C.J. 2005. *Oil Crisis*, Brentwood UK: Multi Science Publishing.

Capling, A., Considine, M. and Crozier, M. 1998. *Australian Politics in the Global Era*, South Melbourne: Longman.

Carr D. 1979. Metropolitan Design in Gentilli J ed. 1979. *Western Landscapes*, University of Western Australia Press, Perth.

Cervero, R. 1996. *Paradigm Shift: From automobility to accessibility planning*, Berkeley: University of California Press.

Cervero, R. 1998. *The Transit Metropolis: A global inquiry*, Washington DC: Island Press.

Cervero R. 2005. Progress in coping with complex urban transport problems in the United States, in *Urban Transport Development, A Complex Issue*, edited by G. Jönson and E. Tengström, Berlin: Springer, 118–133.

Christie, R. 2010. *Independent Public Inquiry into a Long Term Public Transport Plan for Sydney, Final Report*, Sydney. Available at: www.aptnsw.org.au/christie/index.html.

Commonwealth of Australia Constitution Act the Constitution. Available at: http://www.comlaw.gov.au/comlaw/comlaw.nsf/440c19285821b109ca256f3a001d59b7/57dea3835d797364ca256f9d0078c087/$FILE/ConstitutionAct.pdf. [Accessed: 9 April 2007].

Commonwealth Treasury 1997. Budget documents – Horizontal Fiscal Equalisation. Available at: http://www.budget.gov.au/1997-98/horizont.asp. [Accessed: 9 April 2007].

Commonwealth Treasury. 1999. Reform of Commonwealth State Relations. Available at: http://www.treasury.gov.au/documents/196/PDF/round5.pdf. [Accessed: 9 April 2007].

Cresswell, T. 2008. Understanding Mobility Holistically: The Case of Hurricane Katrina, in *The Ehics of Mobilities, Rethinking place, exclusion, freedom and environment*, edited by T. Sager and S. Bergmann. Aldershot: Ashgate, 129–142.

Crozet, Y. 2009. Economic Development and the Role of Travel Time: The Key Concept of Accessibility. *4th International Conference on Future Urban Transport*, April 19th to 21st, Gothenburg, Sweden: Volvo Research and Education Foundations Commissioned Papers, 25–46.

Cumming, David 1995. Petrol price unrest reaches PM level. *Royal Auto*, November 1995, 27

Curtis C. 2008. Planning for Sustainable Accessibility: the implementation challenge. *Transport Policy*, Vol 15 (2)

Curtis, C. 2012. The Challenge of delivering the 'D' in Transit Oriented Development: Examining the Town Planning contribution, *Journal of Transport and Land Use.*

Curtis, C. and Olaru, D. 2010. The relevance of traditional town planning concepts for travel minimisation, *Planning Practice and Research*, 25(1), 49–75.

Curtis C. and Punter J.V. (2004) Design-Led Sustainable Development: The Liveable Neighbourhoods Experiment in Perth, Western Australia. *Town Planning Review* Vol 75 (1)

Curtis, C. and Scheurer, J. 2010. Planning for Sustainable Accessibility: Developing tools to aid discussion and decision-making. *Progress in Planning*, 74, 53–106.

Daly, H. 1996. *Beyond Growth, The economics of sustainable development*, Boston: Beacon Press.

Daly, P. and Hannan, K. 2001. Trains, trams and automobiles, *Royal Auto*, October 2001, 24–25.

Davies, B. and Harré, R. 1990. Positioning: the discursive production of selves. *Journal for the Theory of Social Behaviour*, 201, 43–63.

Davison, G. with Yelland, S. 2004. *Car Wars: How the car won our hearts and conquered our cities*, Crow's Nest Sydney: Allen and Unwin.

De Leuw Cather 1961. *Report on the Geometric Design Study, Sydney Urban Expressways*, San Francisco: De Leuw Cather.

Deffeyes, K.S. 2002. *Hubbert's Peak, The impending world oil shortage*, Princeton, NJ: Princeton University Press.

Denzau, A.T. and Grossman, P. 1993. 'Punctuated equilibria, a model and application of evolutionary economic change', mimeo, Economics Department, Washington University.

Denzau, A.T. and North, D.C. 1994. Shared mental models: ideologies and institutions. *Kyklos*, 471, 3–31.

Department for Transport, Main Roads Western Australia, Ministry for Planning, Fremantle Port Authority, Westrail and Metrobus. 1995. *Metropolitan Transport Strategy*. Perth: Government of Western Australia.

Department of Main Roads, NSW 1976. *The Roadmakers*, Sydney: Department of Main Roads of New South Wales.

Department of Planning and Urban Development. 1990. *Metroplan: a Planning Strategy for the Perth Metropolitan Region*, Perth: Government of Western Australia.

Department of Transport 1996. *State Transport Policy*, Perth: Government of Western Australia.

Department of Transport UK 2007. *Transport Statistics Great Britain, 2007 Edition*, London: Department of Transport Great Minster House, 76 Marsham Street, London SW1P4DR.

Department of Transport UK 2008. *Transport Statistics Great Britain, 2008 Edition*, London: Department of Transport Great Minster House, 76 Marsham Street, London SW1P4DR.

Department of Treasury and Finance 2006. *Statement of Finances 2006/07: Budget Paper No 4*, Department of Treasury and Finance, Melbourne.

Dept. of Transport, Urban Rail Development. 2000. Perth urban rail development project: Supplementary Master Plan, Perth: Department of Transport, Urban Rail Development Office.

DGT [Director General for Transport] 1967. *Rationalisation of Urban Public Transport* Perth: A Review Paper for the Minister for Transport.

DGT [Director General for Transport] 1976. *Public Transport in Perth: A Review Paper for the Minister for Transport by the Director General of Transport* Perth: Office of the Director General of Transport.

DGT [Director General for Transport] 1978. *Development of the Public Transport System in the Perth Metropolitan Region.* Perth: Office of the Director General of Transport.

DGT [Director General of Transport] 1982. *Transport 2000: A Perth study*. Department of Transport, Perth WA.

Dingle T. and Rasmussen, C. 1991. *Vital Connections: Melbourne and its Board of Works 1891 – 1991*; McPhee Gribble: Melbourne.

DoI et al. Department of Infrastructure. 1998a. *Scoresby Transport Corridor Environment Effects Statement – Volume 1* Prepared by Sinclair Knight Merz for VicRoads and the Department of Infrastructure, June 1998.

DoI et al. 1998b. Section 3: Environmental Policy Context, in *Volume G: Scoresby Transport Corridor Environmental Effects Statement* Energy and Planning Research, Sinclair Knight Merz, National Institute for Economic and Industry Research NIEIR, May 1996.

DOTRS. RTA, Sinclair Knight Merz, and PPK 2000. Proposed Western Sydney Orbital Environmental Impact Statement.

Dowling, J. 2011. Cars out in latest Swanston Street revamp, *The Age*, Melbourne, 28th April, 1.

Downs, A. (2004) *Still Stuck in Traffic: Coping with peak hour traffic congestion* Brookings Institution Press

Dudley, G. and Richardson, J. 1998. '"Arenas without rules" and the policy change process: outsider groups and British roads policy'. *Political Studies* 46(4), 727–747.
Dudley, G. and Richardson, J. 1999. Competing advocacy coalitions and the process of frame reflection: a longitudinal analysis of EU steel policy. *Journal of European Public Policy*, 6(2), 225–248.
Dudley, G. and Richardson, J. 2000. *Why Does Policy Change? Lessons from British transport policy 1945–99*, London and New York: Routledge.
Eade, 2000. Issues in Environmental Impact Assessment in Victoria: What Has Scoresby Taught Us? *Urban Policy and Research*, 18(4), 515–27.
EcoTransit, 1993a. Expressways from Everywhere, *Hell on Wheels no. 1*, October 1993.
EcoTransit, 1997. Editorial, A commission of inquiry into the RTA is long overdue, *Hell on Wheels no. 5*, January 1997.
EcoTransit, 2001a. EcoTransit Sydney website. Available at: http://www.ecotransit.org.au/about/ 14/11/2001.
EcoTransit, 2001b. What would another freeway achieve?, *Hell on Wheels no. 7*, February 2001.
EcoTransit, 2001d. The Bay Light Express, *Hell on Wheels no. 7*, February 2001.
Ehrlich, P. 1971. *The Population Bomb*, London: Ballantyne.
Elkington, J. 1997. *Cannibals with Forks: The triple bottom line of 21st Century business*, Oxford: Capstone.
Environmental Protection Act Western Australia [EPA]. 1986. Environmental Impact Assessment Part IV Division 1. Administrative Procedures 2002.
Flyvbjerg, B., Bruzelius, N. and Rothengatter, W. 2003. *Megaprojects and Risk: An anatomy of ambition*, New York: Cambridge University Press.
Flyvbjerg, B., Holm, M.K.S. and Buhl, S.L. 2006. Inaccuracy in traffic forecasts. *Transport Reviews*, 26(1), 1–24.
Galligan, B. and Walsh, C. 1991. *Australian Federalism, Yes or No*, Canberra: Federalism Research Centre, Australian National University.
Garnaut, R. 2008. The *Garnaut Climate Change Review: Final Report*, Port Melbourne, Australia: Cambridge University Press.
Garud, R. and Karnoe, P. eds. 2001a. *Path Dependence and Creation*, Mahwah, New Jersey: Lawrence Erlbaum.
Garud, R. and Karnoe, P. eds. 2001b. Path Creation as a Process of Mindful Deviation, in *Path Dependence and Creation*, edited by Garud, R. and Karnoe, P, Mahwah, New Jersey: Lawrence Erlbaum, 1–42.
German Advisory Council on Global Environmental Change (2005) *World in Transition, Fighting poverty through environmental policy*, London: Earthscan.
Gibson, J.L., Ivancevich, J.M., Donnelly, J.H., Konopaske, R. 2009. *Organizations, Behaviour Structure, Processes*, McGraw-Hill, Boston.
Gilding, P. 2011. *The Great Disruption, How the climate crisis will transform the global economy*, London: Bloomsbury.

Glazebrook, G. 2010. Generating Solutions for Sustainable Transport – the Sydney Experience, Table 1s, Paper given at the GAMUT International Conference, *Sustainable Transport in the Asia-Indo-Pacific, Varied Contexts, Common Aims* June 2nd to 4th.

Gleeson, B.J. and Low, N.P. 2000. *Australian Urban Planning, New challenges, new agendas*, Sydney: Allen and Unwin.

Goodwin, P.B. 1997. *Solving Congestion when we must not build roads, increase spending, lose votes, damage the economy or harm the environment, and will never find equilibrium.* Inaugural Lecture for the Professorship of Transport Policy, University College London.

Goodwin, P.B., Hallett, S., Kenny, F. and Stokes, G. 1991. *Transport: The New Realism*. Report to the Rees Jeffreys Road Fund. Oxford, University of Oxford Transport Studies Unit.

Government of Denmark (2011) *Energy Strategy 2050, From coal, oil and gas to green energy*, Copenhagen: Danish Government.

Government of New South Wales 2005. *City of Cities, A Plan for Sydney's Future*, Sydney: Department of Planning.

Government of New South Wales. 2010. *Metropolitan Transport Plan, Connecting the City of Cities*, Sydney: NSW Department of Transport and Infrastructure.

Government of Victoria. 2002. *Melbourne 2030, Planning for sustainable growth*, Melbourne: Department of Infrastructure.

Government of Victoria. 2006. *Meeting Our Transport Challenges, Connecting Victorian Communities: The Plan*, Melbourne: Victorian Government.

Government of Victoria. 2008. *The Victorian Transport Plan*, Melbourne: Victorian Government.

Government of Victoria. 2009. *Towards an Integrated and Sustainable Transport Future, Policy Statement*, Melbourne: Department of Transport.

Hajer, M. 1997. *The Politics of Environmental Discourse: Ecological modernisation and the policy process.* Oxford: Clarendon Press.

Hall P.A. 1993. Policy paradigms, social learning and the state: the case of economic policy-making in Britain'. *Comparative Politics* 253, 275–294.

Hall, P.A. and Taylor, R.C.R. 1996. Political science and the three new institutionalisms. *Political Studies* 44, 936–957.

Hayek, F. 1944. *The Road to Serfdom*, Sydney: Dymock's Book Arcade.

Heinz, J.P., Laumann, E.O., Nelson, R.L. and Salisbury, R.H. 1993. *The Hollow Core, Private interests in national policy-making*, Cambridge, Mass: Harvard University Press.

Hicks, S., Langoulant, J., Shean, R., and Wauchope, M. 2001. *Government Structures for Better Results: The Report of the Taskforce Established to Review the Machinery of Western Australia's Government.* Perth, Australia: Ministry of the Premier and Cabinet.

Hogarth, Murray 1998a. Outer suburbs in most need of public transport, *Sydney Morning Herald*, 23 February 1998, 4.

Hogarth, Murray 1998b. Clean Sweep, *Sydney Morning Herald*, 24 February 1998, 11.

Hogarth, Murray 1998c. Transport revolution urged to fight smog, *Sydney Morning Herald*, November 3 1998, 5.

Hogarth, Murray 1998d. Labor's pall driven shift, *Sydney Morning Herald*, January 10, 1998,

Holmes, J. 1987. 'Administrative Chronicle: Victoria', Australian Journal of Public Administration, Vol 46(2), 221–224.

Howe, A., Glass, G. and Curtis, C. 2009. Retrofitting TOD and managing the impacts: the case of Subiaco, *in Transit Oriented Development: Making it Happen*, edited by Curtis, C. Renne, J, Bertolini, L. Ashgate: Aldershot.

Imran M. 2010. *Institutional Barriers to Sustainable Transport in Pakistan*, Oxford UK: Oxford University Press.

Isaac, J. 2010. From the President, *Royal Auto*, 78(5), 10.

Johnson, T. 1999. RACV Insight, *Royal Auto*, May 1999, 10.

Keen, S. 2001. *Debunking Economics, The naked emperor of the social sciences*, Annadale New South Wales, Australia: Pluto Press.

Kelly, G.A. 1955. *The Psychology of Personal Constructs*, New York: Norton.

Kelly, P. 1992. *The End of Certainty: the story of the 1980s*, Crow's Nest, Sydney: Allen and Unwin.

Kemp, M. ed (2010) *Zero Carbon Britain 2030, A new energy strategy*, London: New Economics Foundation.

Kemp R., Rip, A. and Schot, J. (2001) 'Constructing transition paths through the management of niches', in *Path Dependence and Creation*, edited by R. Garud, R. and M. Karnoe, New Jersey: Lawrence Erlbaum (Ch 10, pp. 269–301)

Kirman, A. 1989. 'The intrinsic limits of modern economic theory: the emperor has no clothes'. *Economic Journal*, 99, s126-s139.

Klein, S. 2006. *The Secret Pulse of Time, Making sense of life's scarcest commodity*, Melbourne: Scribe Books.

Kopits E. and Cropper M. 2003. 'Traffic fatalities and economic growth', *Policy Research Working Paper, No. 3035* Washington, DC: World Bank.

Krugman P. 1991. 'History and industry location: the case of the manufacturing belt'. *American Economic Review*, 81, 80–83.

Kuhn, T.S. 1962. *The Structure of Scientific Revolutions*, Chicago: Chicago University Press.

Labor Party of Western Australia 2001. *Labor: Getting Transport Back on Track: Pre-election commitments for the 2001 Western Australian State Election* Perth, Western Australia.

Laclau, E. 1994. Why do empty signifiers matter to politics?, in *The Lesser Evil and the Greater Good*, edited by Weeks, J. London: River Orams Press.

Laffin, M. 1987. "No, Permanent Head": Politician-Bureaucrat Relationships in Victoria. *Australian Journal of Public Administration*, 46(2), 37–54.

Laffin, M. 1995. 'The Public Service', in *Reform and Reversal, Lessons from the Coalition Government in New South Wales 1988–1995*, edited by M. Laffin and M. Painter. Melbourne: Macmillan.

Laird, P. and Newman P. 2001. Why it can't go on: the road deficit, in *Back on Track, Rethinking Transport in Australia and New Zealand*, edited by P., Laird, P., Newman, M. Bachels, and J, Kenworthy. Sydney: UNSW Press, 68–90.

Latour, B. 1987. Science in Action, *How to follow scientists and engineers through society, Cambridge*, Mass. Harvard University Press.

Lay, M. 2001. From the President, *Royal Auto* April 2001, 4.

Lay, M. 2003. *Melbourne Miles, The story of Melbourne's roads*, Melbourne: Australian Scholarly Publishing.

Learmonth, Anthony 2001. NRMA follies letter, *Sydney Morning Herald*, 20 March 2001, 13

Legacy, C. 2010, Investigating the knowledge interface between stakeholder engagement and plan-making, *Environment and Planning A*, 42 (11), 2705 –2720.

Liebowitz, S.J. and Margolis, S.E. 1990. The Fable of the Key. *Journal of Law and Economics*, 22, 1–26.

Liebowitz, S.J. and Margolis, S.E. 1995. Path Dependence, Lock-In and History. *Journal of Law, Economics and Organization*, 11(1), 205–226.

Lindblom, C.E. 1959. The science of muddling through. *Public Administration Review.* 19, 79–88.

Lindblom, C.E. 1968. The Policy-Making Process, *Canadian Journal of Political Science*, 4, 295. Englewood Cliffs, N.J.: Prentice-Hall

Lindblom, C.E. 1979. Still muddling, not yet through. *Public Administration Review*, 39, 517–526.

Low, N.P. 1991. *Planning, Politics and the State*, London: Unwin-Hyman.

Low, N.P. 2003. Is Urban Transport Sustainable?, in *Making Urban Transport Sustainable*, edited by N. Low and B. Gleeson. Basingstoke, UK: Palgrave Macmillan, 1–21.

Low, N.P. 2005. The discourse network: a way of understanding policy formation, stability and change in the networked polity, in *The Network Society, A new context for planning*, edited by L. Albrechts and S. Mandelbaum. London and New York: Routledge, 45–56.

Low, N.P. and Astle R. 2009. Path dependence in urban transport: An institutional analysis of urban passenger transport in Melbourne, Australia, 1956–2006. *Transport Policy*, 16(2), 47–58.

Low, N.P. and Gleeson, B.J. 2001. Ecosocialisation or Countermodernisation? Reviewing the shifting 'storylines' of transport planning, *International Journal of Urban and Regional Research* 25(4), 784–803.

Low, N.P. and Power, J.M. 1984. Policy systems in an Australian metropolitan region: political and economic determinants of change in Victoria. *Progress in Planning*, 22(3), 149–215.

Low, N.P., Gleeson, B.J. and Rush, E. 2003. Making believe: institutional and discursive barriers to sustainable transport in two Australian cities.' *International Planning Studies*, 8(2), 93–114.

Lucas, C. 2010a. Plan for hundreds of kilometres of new freeways. *The Age*, 11th Oct, 3.

Lucas, C. 2010b. Brainstorm transport plan could be a road to nowhere. *The Age*, 16th October, 4.

Lucas, C. and Millar, R. 2007. Minister runs down cycle plan. *The Age*, 3rd March, 3.

Lynas, M. 2008. Six Degrees, *Our future on a hotter planet*, Washington: National Geographic.

Mahoney, J. 2000. Path dependence in historical sociology. *Theory and Society*, 29, 507–48.

Main Roads Western Australia, and Department of Transport. 1998. Transform WA: Securing Western Australia's future, Perth, April.

Manning, I. 1991. *The Open Street, Public Transport, Motor Cars and Politics in Australian Cities*, Sydney: Transit Australia Publishing.

March, A. and Low, N. (2005) 'The democratic culture of Victorian urban planning: dilemmas, consent and consensus' in Cryle, D. and Hillier, J. eds *Consent and Consensus, Politics, Media and Governance in Twentieth Century Australia*, Perth, Australia: API Network.

March, J.G. and Olsen, J.P. 1989. *Rediscovering Institutions: the organisational basis of politics*, New York: The Free Press.

March J.G. and Simon, H.A. 1958. *Organizations*, New York: Wiley.

Marchetti, C. 1994. Anthropological invariants in travel behaviour *Technological Forecasting and Social Change*, 47(1), 75–88

McKenzie, Mark 1995a. Blue Skies Forever. *The Open Road*, September/October 1995, 29–30.

McKenzie, Mark 1995b. The air that we breathe – 2000 and beyond. *NRMA Today 1995 Edition 11*, 12–15.

McManus P. 2002. Your Car is as Welcome as You Are: A History of Transportation and Planning in the Perth Metropolitan Region, in *Country: Visions of land and people in Western Australia*, edited by A. Haebich, M. Trinca, A. Gaynor. Western Australian Museum, Perth.

Meadows, D.H., Meadows, D.C., Randers, J. 2005. *The Limits to Growth, the 30 Year Update.* London: Earthscan.

Meadows, D.H., Meadows, D.C., Randers, J. and Behrens, W.W. 1972. *The Limits to Growth.* UK: Universe Books.

Mees P. 2000. *A Very Public Solution, Transport in the dispersed city*. Melbourne: Melbourne University Press.

Mees, P. 2005. 'Responding to the complex transport needs of dispersed wealthy cities: a comparison of Melbourne, Australia and Toronto, Canada' in Jönson, G. and Tengström, E. eds *Urban Transport Development, A complex issue*. Berlin: Springer pp. 32–44.

Mees P. 2009. *Transport for Suburbia, Beyond the automobile age*, London: Earthscan.

Mees, P., Sorupia, E. and Stone, J. 2007. Travel to work in Australian capital cities 1976–2006: an analysis of census data. GAMUT Working Paper.

Metropolitan Region Planning Authority 1971. *Perth Regional Transport Study*, Perth WA: Metropolitan Region Planning Authority.

Metropolitan Transportation Committee 1969. Melbourne *Transportation Study: Volume III* – The Transportation Plan; MTC: Melbourne.

Metz, D. 2008. *The Limits to Travel, How far will you go?* London: Earthscan.

Meyer, M and Miller, E. 2000. *Urban Transportation Planning.*, New York: McGraw Hill second edition.

Milgrom P. and Roberts, J. 1990. The economics of modern manufacturing: technology, strategy and organization. *American Economic Review*, 80, 511–528.

Minister for Transport [Western Australia] 1979. *Urban Public Transport for Perth: Rail and Bus Policy*. Perth: Government of Western Australia.

Ministry of Transport, Victoria 1987. *Melbourne's Arterial Road Strategy: The Next Ten Years*, Melbourne: Ministry of Transport.

Mises, L. 1963. *Human Action, a Treatise on Economics*, Chicago: Contemporary Books.

Mohan D. 2002. Road safety in less-motorized environments: future concerns. *International Journal of Epidemiology*, 31, 527–532.

Moon, J. and Sharman, C. eds 2003. *Australian Politics and Government: the Commonwealth, the states and the territories*, Cambridge, UK, and Port Melbourne: Cambridge University Press.

Moriarty, P. and Honnery, D. (2008) Low mobility: the future for transport. *Futures* 40 (10), 865–872.

Moriarty, P. and Honnery, D. 2010. *Rise and fall of the Carbon Civilisation.* London: Springer.

Morris, L. 1996. Action call to clear the air in divided city. *Sydney Morning Herald*, 25 March 1996, 4.

Morris, W. and Kaufman, J. 1997. Mixed Use development: New designs for New Livelihoods, Brisbane, Queensland Department of Tourism, Small Business and Industry.

Mouritz M., and Ainsworth L. 2009. Successful Delivery Mechanisms: Coordinating Plans, Players and Action, *in Transit Oriented Development: Making it happen*, edited by Curtis, C. Renne, J. and Bertolini, L. Ashgate: Aldershot.

MTT [Metropolitan Transport Trust]. 1961. *Report and Statement of Accounts.* MTT, Perth.

MTT. 1984. *Metropolitan Perth. Passenger Transport Trust Annual Report.* Perth: Metropolitan Transport Trust.

MTT. 1985. *Metropolitan Perth. Passenger Transport Trust Annual Report*, Perth: Metropolitan Transport Trust.

MTT. 1993. *Metropolitan Perth. Passenger Transport Trust Annual Report*, Perth: Metropolitan Transport Trust.

MTT. 1994. *Metropolitan Perth. Passenger Transport Trust Annual Report*, Perth: Metropolitan Transport Trust.

Myer, B. 1993. Metropolitan Strategies for Sydney, in *Spirited Cities: Urban planning, traffic and environmental management in the nineties*, edited by H, Westerman, R. Freestone. Leichhardt, Sydney: The Federation Press.

Nairn, R.J. and Partners. 1990. South-west corridor transport study: final report. Dept. of Transport, Perth, Western Australia.

Nantulya V.M. and Reich M.R. 2002. The neglected epidemic: road traffic injuries in developing countries. *British Medical Journal*, 324(7346), 1139–1141.

Neilsen, G., with Nelson, J.D., Mulley, C., Tegner, G., Lind, G. and Lange, T. 2005. Public Transport – Planning the networks HTrans Best Practice Guide, 2, European Union Interreg IIIB place of publication unspecified.

Neuman, WL. 2003. *Social research methods: qualitative and quantitative approaches*, 2nd Edition. Boston; London: Allyn and Bacon.

Neutze, C.M. 1968. The Process of Urban Development, with particular emphasis on the property market, Paper presented at the Urban Research Unit Seminar, ANU.

Newman undated. The Perth Rail Transformation: Some political lessons learned. Available at : http://sustainability.curtin.edu.au/local/docs/The_Perth_Rail_Transformation.pdf. [Accessed: 8 November 2011].

Newman P. and Kenworthy, J. 1999. Sustainability and Cities, Overcoming Automobile Dependence. Washington DC: Island Press.

Newman, P.W.G. and Kenworthy, J.R. 1989. Cities and Automobile Dependence: A Sourcebook, Aldershot, UK: Gower Technical

Nielsen, K, Jessop, B and Hausner, J. 1995. Strategic Choice and Path Dependency in Post-socialism Cheltenham: Edward Elgar.

North, D. 1990. Institutions, Institutional Change and Economic Performance, Cambridge: Cambridge University Press

Norton, P. 1996. Fighting Traffic, US urban transportation policy and urban congestion 1955–1970, Charlottesville: University of Virginia.

NRMA, 1995. Our Biggest Pollution Problem. *The Open Road*, July/August 1995, 15–16.

NRMA, 1995a. Sydney and its transport beyond 2000 NRMA discussion paper prepared in response to the RTA's Draft State Road Network Strategy and the State Rail Authority's CityRail Strategic Plan 1994–2016.

NRMA, 1996. Shaping Sydney's Transport – a framework for reform Discussion paper for Clean Air Task Force, prepared by NRMA Clean Air 2000 Project Team.

NRMA, 1998. Clean Air 2000 Strategic Action Plan: Objectives, Targets and Strategies 1997–2000.

NRMA, 2001. NRMA Annual Report [Accessed: 19 October 2001].

NRMA, 2004. NRMA website: Demutualisation. Available at: http://www.mynrma.com.au/member_centre/company_profile/demutualisation/index.shtml. [Accessed: 25 February 2004].

NSW Department of Environment and Planning. 1988. *Sydney into Its Third Century: Metropolitan Strategy for the Sydney Region*, Sydney: NSWDoP.

NSW Department of Planning. 1993. *Cities For The 21st Century: Integrated urban management for Sydney, Newcastle, the Central Coast and Wollongong*, Sydney: NSWDoP.

NSW Department of Planning. 1993. *Sydney's Future: A discussion paper on planning the greater metropolitan region*, Sydney: NSWDoP.

NSW Department of Transport. 1974. A Transport Plan for Sydney – 2000: Sydney area transportation study May 1974, Sydney: DoT.

NSW Department of Transport. 1993. Integrated Transport Strategy for the Greater Metropolitan Region, Sydney: DoT.

NSW Government. 1997. A Framework for Growth and Change: The review of strategic planning the greater metropolitan region, Sydney: NSW Government.

OECD 2008. Greenhouse Gas Reduction Strategies in the Transport Sector, Preliminary Report, Organization for Economic Cooperation and Development. International Transport Forum. Available at: www.internationaltransportforum.org/Pub/pdf/o8GHG.pdf. [Accessed: 28 April 2011].

Owens 1995. From "predict and provide" to "predict and prevent"?: Pricing and planning in transport policy. *Transport Policy*, 2(1), 43–49.

Oxford English Dictionary. 2006. Oxford University Press. Oxford. Available at: http://www.oed.com/. [Accessed: 15 June 2006.

Parliament of Victoria 2010. *Transport Integration Act, 2010*, Melbourne.

Patton, MQ. 2002. Qualitative research and evaluation methods, 3rd Edition, Sage Publications, Thousand Oaks, California.

Perth City Rail Advisory Committee. 2002. Report of the Perth City Rail Advisory Committee to the Minister for Planning and Infrastructure, Hon Alannah MacTiernan MLA, Perth: Perth City Rail Advisory Committee.

Peters, G. 1999. Institutional Theory in Political Science: The 'New Institutionalism', London and New York: Pinter.

Pierson, P. 2000a. Increasing returns, path dependence, and the study of politics. *American Political Science Review*, 94(2), 251–267.

Pierson, P. 2000b. Not just what but When: Timing and Sequence in Political Processes. *Studies in American Political Development*, 14, 72–92.

Pierson, P. 2004. *Politics in Time, History, institutions and social analysis*, Princeton and Oxford: Princeton University Press.

Powell, T. 2003. The 1960s, Planning for growth and chang' in *Planning or Politics, Town planning for the Sydney region since 1945*, edited by J. Toon and J. Falk. Sydney: Planning Research Centre, University of Sydney, 65–90.

PTUA. 1991. Greening Melbourne with Public Transport Public Transport Users Association, Melbourne.

PTUA. 1995. Wrong Way Go Back: The Alternative to Melbourne's Freeway Explosion Public Transport Users Association, Melbourne.

PTUA. 1997. Making Connections: Environmentally Friendly Transport for Melbourne's Outer East Public Transport Users Association, Melbourne.

PTUA. 2001. *Who We Are*. Public Transport Users Association website. Available at: http://home.vicnet.net.au/~ptua/. [Accessed: 14 November 2001].

RACV. 1980a. Over 20% of car running costs are Government taxes. *Royal Auto,* May 1980, 2.

RACV. 1980b. AAA seeks $900 million for roads. *Royal Auto*, February 1980, 3.

RACV. 1980c. Support for AAA road funds campaign. *Royal Auto*, February 1980, 4, 16–17.

RACV. 1980d. The possible petrol substitute. *Royal Auto*, March 1980, 9–10.

RACV. 1981. "We've paid enough" in petrol taxes. *Royal Auto*, July 1981, 2.

RACV. 1981a. Government charges increase car costs. *Royal Auto*, November 1981, 2.

RACV, 1987. Petrol Pricing – What's really going on? *Royal Auto*, May 1987, 4.

RACV, 1987a. Road Funding. Crisis. *Royal Auto*, June 1987, 5.

RACV, 1995. Petrol pricing. *Royal Auto* August 1995, 3.

RACV, 2001. About RACV, Available at: http://www.racv.com.au/shop/about.asp. [Accessed: 19th October 2001].

Rattray, A. and Sinclair, A. 1978. *A Discussion Paper of Transport Planning in Sydney to 1975*, Canberra: Bureau of Transport Economics.

Roe7 Alliance. 2004. Roe Highway Stage 7 Extension South Street, Kwinana Freeway.: Assessment number 1466. Report on issues from submissions and responses to issues. Available on CD Rom in EPA Bulletin 1138.

Rogers, E. 1962. *Diffusion of Innovations*, New York: Free Press.

Romer, P.M. 1986. Increasing returns and long run growth, *Journal of Political Economy*, 94, 1002–1037.

Romer, P.M. 1990. Are non-convexities important to understanding growth? *American Economic Review*, 80, 97–103.

Roth, G. 1967. *Paying for Roads, The economics of traffic congestion*, Harmondsworth, UK, Penguin Books.

Rothengatter, W. 2000. Obstacles to the use of economic instruments in transport policy, *Internalising the Social Costs of Transport*. Chapter 5. Available at: http://www.internationaltransportforum.org/IntOrg/ecmt/taxes/pdf/94socCosts6e.pdf. Paris: OECD/ECMT.

Rothman, J. 1974. *Planning and Organizing for Social Change*, New York: Columbia University Press.

Royal Automobile Club of Western Australia. 1962a. the Federal Budget A.A.A. Roads Case. *The Road Patrol*, 21.

Royal Automobile Club of Western Australia. 1962b. World oil reserves still increasing. *The Road Patrol*, 11.

Royal Automobile Club of Western Australia. 1964a. Living Room Motor Vehicle. *The Road Patrol*, 21.

Royal Automobile Club of Western Australia. 1964b. Traffic in Towns THE BUCHANAN REPORT – A Review by Steve Blatchford. *The Road Patrol*, 26, 7, 30.

Royal Automobile Club of Western Australia. 1966. Editorial – Rights of the motorist. *The Road Patrol*, 4.

Royal Automobile Club of Western Australia. 1968a. Aust. Freeway Building Lags. *The Road Patrol*, 3.

Royal Automobile Club of Western Australia. 1970a. Car Maker on Pollution. *The Road Patrol*, 29.

Royal Automobile Club of Western Australia. 1970b. Pollution and the motor car, *The Road Patrol*, 4.

Royal Automobile Club of Western Australia. 1971. Automobile bodies campaign against more tax slugs, *The Road Patrol*, 1.

Royal Automobile Club of Western Australia. 1973a. Is Freeway such a dirty word? *The Road Patrol*, 8.

Royal Automobile Club of Western Australia. 1973b. Modern freeways a blessing not a curse, *The Road Patrol*, 2.

Royal Automobile Club of Western Australia. 1978. Car is here to stay, *The Road Patrol*, 2.

Royal Automobile Club of Western Australia. 1979. Oil supply "limited and irreplaceable" RAC submission on energy use, *The Road Patrol*, 1.

Royal Automobile Club of Western Australia. 1980. Many back move on roads funding, *The Road Patrol*, 1.

Royal Automobile Club of Western Australia. 1982. The RAC view – Road system jeopardy, *The Road Patrol*, 2.

Royal Automobile Club of Western Australia. 1988. Streetwise – The sound of money, *The Road Patrol*, 36.

Royal Automobile Club of Western Australia. 1990a. The RAC view – On the environment, *The Road Patrol*, 2.

Royal Automobile Club of Western Australia. 1990b. The RAC view – Road funding – Party policies, *The Road Patrol*, 2.

Royal Automobile Club of Western Australia. 1991a. Annual Report, *The Road Patrol*, 8.

Royal Automobile Club of Western Australia. 1991b. Drive GREEN, *The Road Patrol*, 4.

Royal Automobile Club of Western Australia. 1991c. RAC's environmental strategy, *The Road Patrol*, 5.

Royal Automobile Club of Western Australia. 1994a. Stop The Carnage – Fix the Roads, *The Road Patrol*, 61.

Royal Automobile Club of Western Australia. 1994b. Stop The Carnage – Fix the Roads, *The Road Patrol*, 65.

Royal Automobile Club of Western Australia. 1995a. Fix Australia. – Fix the Roads 1995 summit, *The Road Patrol*, 26.

Royal Automobile Club of Western Australia. 1995b. Smoky cars campaign takes off, *The Road Patrol*, 20.

Royal Automobile Club of Western Australia. 1996. Motorists do their bit for World Environment Day, *The Road Patrol*, 20.

Royal Automobile Club of Western Australia. 2000a. *The Road Patrol*, 5.

Royal Automobile Club of Western Australia. 2000b. More trains and buses the key to cleaner air, *The Road Patrol*, 30.

Royal Automobile Club of Western Australia. 2000c. The RAC's role, *The Road Patrol*, 9.

Royal Automobile Club of Western Australia. 2000d. RAC air care – raising awareness, *The Road Patrol*, 14–5.

Royal Automobile Club of Western Australia. 2000e. Take a deep breath – maybe not! *The Road Patrol*, 26.

Royal Automobile Club of Western Australia. 2002a. Planting the seeds for the future, *The Road Patrol*, 12.

Royal Automobile Club of Western Australia. 2002b. State of our roads, *The Road Patrol*, 6–7.

Royal Automobile Club of Western Australia. 2003. Motorists expect better from Budget, *The Road Patrol*, 31.

Royal Automobile Club of Western Australia. 2005/6b. Keeping WA's wheels turning, *The Road Patrol*, 9–11.

RTA, undated. The Western Sydney Orbital Overview Report: Starting the environmental assessment.

Russell, B and White, N. (2009) *Federal Ingredients in Transportation Policy: Australia, USA, Germany*, GAMUT Working Paper, Melbourne: Australasian Centre for Governance and Management of Urban Transport, The University of Melbourne.

Ryan, M. 2008. Some greenhouse-friendly tax reforms. *Tax Policy Journal*, 4.

Sabatier, P. A. (1999) *Theories of the Policy Process*, Boulder, Colorado: Westview Press.

Sabatier, P.A. and Jenkins-Smith, H.C. 1993. *Policy Change and Learning, An advocacy coalition approach*, Boulder, USA: Westview Press.

SACTRA [Standing Advisory Committee on Trunk Road Assessment] 1994. *Trunk Roads and the Generation of Traffic*, London: HMSO.

Sandercock, L. 1975. *Cities for Sale*, Melbourne: Melbourne University Press.

Sandercock, L. 1998. *Towards Cosmopolis, Planning for multicultural cities*, Chichester, UK, Wiley.

Schaeffer, K. H. and Sclar, E. 1975. *Access for All. Transportation and Urban Growth.* Baltimore, USA, Penguin Books.

Searle, G. 1997. Sydney's recent transport planning: under economic ascendancy, paper presented to 15th Pacific Regional Science Conference Organisation, Wellington, New Zealand, December.

Self, P. 1993. *Government by the Market?* Basingstoke UK: Macmillan.

Self, P. (2000) *Rolling Back the Market, Economic dogma and political choice*, Basingstoke UK: Macmillan.

Selmes, 1995. MTA backs Whitlam' Sydney Morning Herald Motoring Section, 28 September 1995, 29.

Shafritz, J.M., Russell, E. and Borick, C.P. 2009. Introducing Public Administration sixth edition. New York: Pearson Longman.

Simmie, J. (1974) *Citizens in Conflict, The sociology of town planning*, London: Hutchinson.

Smiles, S. 2007. Road to riches bypasses Victoria, State claims. *The Age*, 4.

Solomon, S. et al. 2007. Climate Change 2007, Geneva: Intergovernmental Panel on Climate Change.

South West Corridor Rapid Transit Steering Committee. 1990. Rapid Transit in the South West Corridor – Draft Report. South West Corridor Rapid Transit Steering Committee. Perth, Western Australia.

South West Metropolitan Transit Planning Steering Committee. 1999. South West Metropolitan Railway: Perth to Jandakot, Rockingham and Mandurah Master Plan, South West Metropolitan Railway Master Plan. Western Australia.

Spearritt, P. and De Marco, C. 1988. *Planning Sydney's Future*, Sydney: Allen and Unwin.

Sraffa, P. 1926. The law of returns under competitive conditions. *Economic Journal*, 40, 538–550.

State Planning Authority of NSW 1967. *Sydney Region, Growth and Change: Prelude to a plan*, Sydney.

State Planning Authority of NSW 1968. *Sydney Region: Outline Plan* 1970–2000, Sydney.

Steering Committee of the South West Corridor Transport Study. 1990. A Rapid Transit System to Rockingham and Mandurah? A preliminary discussion paper on the feasibility of developing a rapid transit public transport. system in Perth's south west corridor. Perth: Department for Planning and Urban Development.

Steering Group and Working Group Buchanan Report. 1963. *Traffic in Towns, A study of the long term problems of traffic in urban areas,* Reports of the Steering and Working Groups appointed by the Minister of Transport, London: HMSO.

Stephenson G. 1992. *On a human scale: a life in urban design.* South Fremantle, Western Australia, Fremantle Arts Centre Press.

Stephenson, G. and Hepburn, A. 1955. *Plan for the Metropolitan Region Perth and Fremantle*. Perth, WA: Government Printing Office.

Stern, N.H. 2007. *The Economics of Climate Change*: The Stern Review, Cambridge: Cambridge University Press.

Stone, J.A. 2008. *Political Factors in the Rebuilding of Mass Transit: An investigation of failure in Melbourne since 1970 through comparisons with Perth and Vancouver*, PhD Thesis Faculty of Life and Social Sciences, Swinburne University of Technology, Melbourne, Australia. Available at: http://hdl.handle.net/1959.3/36049.

Stone, J.A. 2009. Contrasts in reform: how the Cain and Burke years shaped public transport in Melbourne and Perth. *Urban Policy and Research*, 27(4), 419–434.

Stretton, H. 1969. *The Political Sciences, General principles of selection in social science and history*, New York: Basic Books.

Stretton, H. 1975. *Interests, Ideas for Australian Cities*, Melbourne: Georgian House.

Texas Transportation Institute 2010. *Urban Mobility Report 2010*, Table 2 'What Congestion Means to Your Town', Available at: http://mobility.tamce.edu.au/ums/. [Accessed: 27th January 2010].

Toon, J. and Falk, J. 2003. *Planning or Politics, Town planning for the Sydney region since 1945*, Sydney: Planning Research Centre, University of Sydney.

Torfing, J. 1998. *Politics, Regulation and the Modern Welfare State*, Basingstoke: Macmillan.

Torfing, J. 2001. Path-dependent Danish welfare reforms: the contribution of the new institutionalisms to understanding evolutionary change. *Scandinavian Political Studies*, 24(4), 277–309.

[TRAC] Transport Action Coalition. 1997a. Newsletter August 1997.

[TRAC] Transport Action Coalition. 1997b. Newsletter October 1997.

[TRAC] Transport Action Coalition. 1998a. Newsletter March 1998.

[TRAC] Transport Action Coalition. 1998b. Newsletter October 1998.

[TRAC] Transport Action Coalition. 1999. Newsletter February – March 1999.

[TRAC] Transport Action Coalition. 2000. Newsletter October 2000.

Transperth and Western Australia. South West Area Transit Steering Committee. 1992. South west area transit, SWAT: technical information pack. Transperth, East Perth, W.A.

Transperth and Department of Transport WA. 1993. *Metropolitan Transport Strategy–Perth: A More Responsive and Integrated Public Transport Service for Perth*. Perth: Department of Transport.

Transport. 1998. *Better Public Transport: Ten Year Plan for Transperth 1998–2007*, Perth: Government of Western Australia.

Tranter, P.J. 2010. Speed Kills: The complex links between transport, lack of time and urban health. *Journal of Urban Health*, 87(2), 155–164.

Travers Morgan Pty. Ltd, Sinclair Knight Buchanan, Western Australia. South West Metropolitan Region Transport Planning Committee. 1992. South West Metropolitan Region transport planning study: study report. The Committee. Perth, Western Australia.

UNFCCC. 2011. *The Kyoto Protocol*. Available at: http://unfccc.int/essential_background/kyoto_protocol/items/1678.php, United Nations Framework Convention on Climate Change.

Vaile, M. 2007. *The Road to AusLink 2*, Speech VS04/2007 ABARE Outlook Conference Canberra 7th March 2007. Available at: http://www.ministers.dotars.gov.au/mv/speeches/2007/VS04_2007.htm.

VicRoads. 1994. *Linking Melbourne*. Strategic Transport Planning Department, VicRoads: Melbourne.

Victoria, Minister's second reading speech, Victorian Parliamentary Debates 5th May 1983, 4324.

Victoria, Ministry of Transport 1987. *Melbourne's Arterial Road Strategy: The Next Ten Years*, Government of Victoria, Ministry of Transport, Melbourne.

Victoria, Ministry of Transport 1988. *MetPlan: Metropolitan Public Transport Industry Plan*; Government of Victoria, Ministry of Transport, Melbourne.

Victorian Transport Study 1980a. *Final Report*, Government of Victoria, Ministry of Transport, Melbourne.

Victorian Transport Study 1980b. *Background to Transport Finance*, Government of Victoria, Ministry of Transport, Melbourne.

Vigar, G. 2002. *The Politics of Mobility, Transport, the environment and public policy* London: Spon.

Wainwright. 1998. Schemes and Dreams. *Sydney Morning Herald*, 28 November 1998, 37

Wainwright, Robert and Totaro, Paola 1998. How to squeeze the cars out of Sydney: $5 billion solution to transport chaos. *Sydney Morning Herald*, 24 March 1998, 1.

Walton, Julie 2000. Ticket to ride. *Sydney Morning Herald*, 15 May 2000, 13.

Ward, P. 2007. *Under a Green Sky, Global warming, the mass extinctions of the past, and what they can tell us about our future*, New York: Harper-Collins.

Wardlaw, H. 2003. The 1950s, Reality Strikes the County of Cumberland in *Planning or Politics, Town planning for the Sydney region since 1945*, edited by J. Toon and J. Falk. Sydney: Planning Research Centre, University of Sydney, 49–64.

Weber, M. [1919]. 1946. Politics as a Vocation, in *Max Weber – Essays in Sociology*, edited by H.H. Gerth, and C.W. Mills. New York: Oxford University Press.

Weber, M. 1946. *Max Weber: Essays in Sociology*, Gerth, H.H. and Mills, C.W., eds., New York: Oxford University Press.

Western Australia, Parliamentary Debates, Legislative Assembly, 20 June 2002, 11820d Ms Alannah MacTiernan, Armadale – Minister for Planning and Infrastructure.

Western Australia, Parliamentary Debates, Legislative Assembly, 20 September 2005, 5603b Ms Alannah MacTiernan, Armadale – Minister for Planning and Infrastructure.

Western Australia, Parliamentary Debates, Legislative Assembly, 31 July 2001, 1695c–1706a Dr Geoff Gallop, Victoria Park – Premier.

Western Australian Government Railways Commission. 1964. Annual Report, Perth: Government of Western Australia.

Western Australian Planning Commission. 2000. Liveable Neighbourhoods – Street Layout, Design and Traffic Management Guidelines. Western Australian Planning Commission, Perth.

Western Australian Planning Commission WAPC, 2004. Network City: Community Planning Strategy for Perth and Peel, Perth: WAPC.

Western Australian Planning Commission WAPC. 2005. Network City—A Milestone in Metropolitan Planning. Statement by the WAPC on Partnerships for Planning Perth and Peel, Perth: WAPC.

Western Australian Planning Commission WAPC, 2009. Directions 2031: A Draft Spatial Framework for Perth and Peel, Perth: WAPC.

WHO [World Health Organisation] 2004. *World Report on Road Traffic Injury Prevention* Peden M, Scurfield R, Sleet D, Mohan, D, Hyder A.A, Jarawan E, Mathers C Eds. Geneva: World Health Organization.

Wildavsky, A. (1973). If planning is everything, maybe it's nothing. *Policy Sciences*, 4(2), 127–153.

Wilson, J., Thompson, J. and McMahon, A. 1996. *The Australian Welfare State*, Melbourne: Macmillan.

Winston, D. 1957. *Sydney's Great Experiment: The progress of the Cumberland County Plan*, Sydney: Angus and Robertson.

Woodcock, J., Edwards P., Tonne, C., Armstrong B.G., Ashiru, O, Banister, D., Beevers, S., Chowdhury, Z., Cohen, A., Franco, O.H., Haines, A., Hickman, R., Lindsay, G., Mohan, D., Tiwari, G., Woodward, A. and Roberts, I. 2009. Public health benefits of strategies to reduce greenhouse-gas emissions: urban land transport. *The Lancet*, 374, 1930–1943.

Woodlief, A. 1998. The path-dependent city. *Urban Affairs Review*, 33(3), 405–437. Available at: www.nrma.com.au/media/abt-2001annualrep-roadservice.pdf.

Yago, G. 1984. *The Decline of Transit, Urban transportation in German and US Cities, 1900–1970*, Cambridge: Cambridge University Press.

Yiftachel, O. and Kenworthy, J. 1992. The planning of metropolitan Perth: Some critical observations, in *Urban and Regional Planning in Western Australia: Historical and Critical Perspectives*, edited by D. Hedgcock and O. Yiftachel. Perth: Paradigm Press, 131–148.

Young, K. and Mills, L. 1980. *Public Policy Research: A review of qualitative methods*, London: Social Science Research Council.

Zahavi, Y. and Talvitie, A. 1980. Regularities in travel time and money expenditures. *Transportation Research Record*, 750, 13–19.

Zeibots, Michelle 1997. 'Breeding gridlock: Induced traffic – what it is and how it happens' Hell on Wheels no. 5, January 1997.

Zhukov, Greg 1997. Eastern Distributor "fraud". *Hell on Wheels no. 5*, January 1997,

Appendix 1: Roads Policy Discourse – Documents analysed (Chronological)

Melbourne

Melbourne and Metropolitan Board of Works, (1954). Melbourne Metropolitan Planning Scheme, Melbourne: Melbourne and Metropolitan Board of Works.

Melbourne Transportation Committee (1969) Melbourne Transportation Study Vol 3 The Transportation Plan. Melbourne: Metropolitan Transport Committee

Melbourne and Metropolitan Board Of Works (1971) Planning Policies for The Melbourne Metropolitan Region. Melbourne: Melbourne and Metropolitan Board of Works.

Melbourne and Metropolitan Board of Works (1974) Report On General Concept Objections 1974 Planning policies for the Melbourne and metropolitan region and amending planning schemes 3 &31'. Melbourne: Melbourne and Metropolitan Board of Works

State Co-ordination Council (1978) Transport Plan 1978 for Victoria. Volume 1. The Strategy. Melbourne: State Co-ordination Council

Melbourne and Metropolitan Board of Works (1979) Alternative Strategies For Metropolitan Melbourne. Melbourne: Melbourne and Metropolitan Board of Works

Melbourne and Metropolitan Board of Works (1980) Metropolitan Strategy. Melbourne: Melbourne and Metropolitan Board of Works

Lonie, W.M. (1980). Victoria Transport Study: Metropolitan Roads. Melbourne: [S.L.]

Melbourne and Metropolitan Board of Works (1981) Metropolitan Strategy Implementation. Melbourne: Melbourne and Metropolitan Board of Works

Ministry of Transport, Victoria (1983). Transport Development. Policies and Processes. Melbourne: Government of Victoria

Ministry of Transport (1986). Metropolitan Arterial Road Access Study. Strategy Development Final Report M4. Melbourne: Government of Victoria.

Department of Planning and Environment (1987) Shaping Melbourne's Future. The Government's Metropolitan Policy. Melbourne: Government Printer.

Ministry for Planning and Environment, June, 1989. The Greenhouse Challenge: The Victorian Government's Response – A draft strategy for public comment. Melbourne: Ministry for Planning and Environment.

Department of Planning and Housing (1992) A Place to Live: Shaping Victoria's Future, Melbourne: Government Printer

Department of Planning and Development, Department of Transport, Department Of Conservation and Natural Resources (1994) Melbourne Metropolitan Strategy: Discussion Paper, Melbourne: Department of Planning and Development

Government of Victoria (1995) Living Suburbs: A policy for metropolitan Melbourne into the 21st century. Melbourne: Government of Victoria

Department of Infrastructure (1996). Transporting Melbourne, a strategic framework for an integrated transport system in Melbourne. Government of Victoria: Melbourne.

Thwaites, J. (1999). State Planning Agenda: A Sensible Balance. Melbourne: Minister for Planning, Department of Infrastructure.

Department of Infrastructure (2000) Challenge Melbourne: Issues in Metropolitan Planning for the 21st Century

Sydney

State Planning Authority (1967) Growth and Change: Prelude to a Plan, Sydney: State Planning Authority of NSW.

State Planning Authority (1968) Sydney Region Outline Plan 1970–2000, Sydney: State Planning Authority.

Anon. (1974) Sydney Area Transportation Study Volume 3: Passenger Transport Systems, Sydney: n.p.

Urban Transport Advisory of New South Wales (1976) A Report to the Minister for Transport and Highways and Minister for Planning and Environment, Sydney: Urban Transport Advisory Committee of New South Wales

Urban Transport Advisory Committee of New South Wales (1979) A Report to the Minister for Transport, Sydney: Urban Transport Advisory Committee of New South Wales.

NSW Planning and Environment Commission (1980) Review Sydney Region Outline Plan, Sydney: NSW Planning and Environment Commission.

Department of Main Roads (1987) Roads 2000 – Sydney Region, Sydney: Department of Main Roads.

NSW Department of Environment and Planning (1988) Sydney into its Third Century: Metropolitan Strategy for the Sydney Region, Sydney: NSW Department of environment and Planning.

Department of Planning (1993) Sydney's Future: A Discussion Paper on Planning the Greater Metropolitan Region, Sydney: Sydney: Department of Planning.

Department of Planning (1995a) Cities for the 21st Century, Sydney: Sydney: Department of Planning.

Department of Planning (1995b) Integrated Transport Strategy for the Greater Metropolitan Region, Sydney: Sydney: Department of Planning.

Review of Strategic Planning, the Greater Metropolitan Region (1997) A Framework for Growth and Change Sydney: The Review of Strategic Planning, the Greater Metropolitan Region.

NSW Department of Urban Affairs and Planning (1998) Shaping Our Cities, Sydney: NSW Department of Urban Affairs and Planning

New South Wales Government (1998) Action for Transport 2010: An Integrated Transport Plan for Sydney, Sydney: New South Wales Government.

Perth

Stephenson, G. and Hepburn, A. (1955) *Plan for the Metropolitan Region Perth and Fremantle,* Perth: Government Printing Office.

Town Planning Department. (1962) *Metropolitan Region Scheme Report*, Perth: The Metropolitan Region Planning Authority.

The Metropolitan Region Planning Authority. (1968) *Perth: Region and People*, Perth: The Metropolitan Region Planning Authority.

Ritter, P. (1968) *City Planning Perth 1965–1967*, Perth: Planned Environment and Education Research.

The Metropolitan Region Planning Authority. (1970) *Perth Regional Transport Study*, Perth: The Metropolitan Region Planning Authority.

The Metropolitan Region Planning Authority. (1970) *The Corridor Plan for Perth*, Perth: The Metropolitan Region Planning Authority.

The Metropolitan Region Planning Authority. (1972) *Report on the Corridor Plan for Perth*, Perth: The Metropolitan Region Planning Authority.

Director General of Transport. (1982) *Transport 2000: A Perth study*, Perth: Transport.

Department of Planning and Urban Development. (1990) *Metroplan: a Planning Strategy for the Perth Metropolitan Region*, Perth: Government of Western Australia.

Department of Planning and Urban Development, Department of Transport, Main Roads Department and Transperth. (1991) *Road Reserves Review Final Report, Perth, Perth:* Government of Western Australia.

Department for Transport, Main Roads Western Australia, Ministry for Planning, Fremantle Port Authority, Westrail and Metrobus. (1995) *Metropolitan Transport Strategy*. Perth: Government of Western Australia.

Department of Transport. (1996) *State Transport Policy*, Perth: Government of Western Australia.

Ministry for Planning. (1997) *State Planning Strategy,* Perth: Western Australian Planning Commission.

Appendix 2: Public Transport Policy Discourse – Documents analysed (Chronological)

Melbourne

MMTB, 1965–1982 MMTB Report and Statement of Accounts 1965–1982 [18 reports]

MTA, 1984–1989 Metropolitan Transit Authority Annual Report 1984–1989 [6 reports]

MTA, 1988a MetPlan: Metropolitan Public Transport Industry Plan

PTC, 1990–1999 Public Transport Corporation Annual Report 1990–1999 [10 reports]

VR, 1965–1982 VicRail Annual Report 1965–1982 [18 reports] [exact titles varied]

Sydney

Commissioner for Government Transport New South Wales 1960, Report of the Commissioner for Government Transport, Sydney: Department of Government Transport.

Commissioner for Government Transport New South Wales 1965, Report of the Commissioner for Government Transport, Sydney: Department of Government Transport.

Commissioner for Department Railways 1972, Annual Report of Commissioner for Department Railways.

Commissioner for Government Transport New South Wales 1972, Report of the Commissioner for Government Transport, Sydney: Department of Government Transport.

Public Transport Commission 1973, Report of the Bus Task Force of the Public Transport Commission of New South Wales.

Urban Transit Authority of New South Wales 1981, Annual Report.

State Transit Authority of New South Wales 1989, Annual Report.

State Rail Authority of New South Wales 1990, Annual Report.

NSW Department of Planning 1993, Integrated transport strategy for Greater Sydney: a first release for public discussion.

State Rail Authority of New South Wales 1995, Annual Report.

State Transit Authority of New South Wales 1995, Annual Report.
Department of Transport 1998, Action for transport 2010: an integrated transport plan for Sydney.

Perth

Metropolitan Transport Trust (MTT) Report & Statement of Accounts – 1959–1985.
Western Australian Government Railways Commission 1964, 'Annual Report'.
Office of the Director General of Transport 1967, 'Rationalisation of Urban Public Transport'
Office of the Director General of Transport [ODGT] 1976, 'Public Transport in Perth: A Review Paper for the Minister for Transport by the Director General of Transport'.
Director General of Transport [DGT] 1978, 'Development of the Public Transport System in the Perth Metropolitan Region'.
Minister for Transport 1979, ' Urban Public Transport for Perth: Rail and Bus Policy'.
Transperth 1986, Annual Report.
Transperth 1991, Annual Report.
Transport 1998, 'Better Public Transport : Ten Year Plan for Transperth 1998–2007'

Appendix 3

Storyline	Question #	Question
		The Engineer's Tale
Predict & Provide	Q1	The best way to plan a transport system is to base future provision of infrastructure on accurate projections of travel demand.
Predict & Provide	Q2	Carefully calibrated statistical modelling based on accurate survey data is the best available tool for planning a transport system.
Predict & Provide	Q3	It is important to provide for high levels of mobility.
Free Movement	Q4	The best urban transport system is one in which people are able to move freely and quickly around the system.
Free Movement	Q5	Traffic jams on roads are a sign that a city's transport system is not functioning optimally.
Free Movement	Q6	The most important goal today in transport planning is to reduce the time spent in movement of freight.
Balanced Transport	Q7a	Within an urban system there should be a balance of investment in the infrastructure required for different transport modes (roads, public transport, cycling, walking).
Balanced Transport	Q7b	Within an urban system there should be an approximately equal balance of investment between I) roads, and 2) public transport and other modes.
Bypass	Q12	In [city] investment in improved roads is required to bypass congested areas.
Bottleneck	Q13	Investment in improved road infrastructure is needed to relieve 'bottlenecks' in the road systems that cause congestion.
Missing links	Q15	Ideally a city should have a uniformly good road system throughout the urban area in which there should be no 'missing links' in the road network.
Technofix	Q16	Even if private vehicles are shown to be environmentally damaging, there will always be a need to place a high priority on improving the road system because new technologies will develop to allow road vehicles to become environmentally friendly (non-polluting, near zero greenhouse gas emissions, not dependent on fossil fuels etc.).
Technofix	Q17	There will always be a need to place a high priority on improving the road system because new technologies will develop to prevent congestion occurring from road use (road pricing, on-road vehicle control technologies etc).
Safety	Q51	It is important to invest in new road building to avoid major safety issues on existing roads.

		The Economist's Tale
Public transport deficit	Q54	In today's economic climate of globalization and competition for scarce government resources governments cannot afford to sustain a high level of public transport deficit.
Public transport = welfare	Q32	Public transport is best regarded as a form of welfare provision for people who cannot, for various reasons, travel by car.
Modern Lifestyle	Q43	Car-dependent cities are essentially the result of the lifestyle choices of their citizens.
Travel as opportunity	Q18	All travel in an urban system occurs because people and firms want to maximize their opportunities to undertake desired activities in spatially separated locations (for example between home, work, recreation etc.).
Social rationality -Aggregate of indiv. decisions	Q25	The transport system in cities should reflect the aggregate of the travel choices of individuals.
People's Choice	Q44	The construction of new or enlarged primary distributor roads in [city] is a response to public demand.
People's Choice	Q45	Continued investment in roads is necessary because people will not change their travel behaviour.
Market solutions best	Q20	The best solutions to transport management problems are those in which individuals are allowed to make their own travel choices within their budgets.
Market solutions best	Q21	Congestion pricing is an efficient way of constraining travel demand to fit a given level of transport infrastructure provision (e.g. the congestion charge recently introduced in central London or other more refined methods.)
Market solutions best	Q22	Congestion pricing is a fair way of constraining travel demand to fit a given level of transport infrastructure provision.
Private transport = efficiency	Q19	Transport is most efficiently provided by private companies, operating under competitive conditions in free markets.
Time saving = cost saving	Q14	Time saved in travel can reasonably be equated with cost savings for firms.
Consumer surplus	Q23	The monetary benefit to consumers of transport in a region can be derived from the time saved in individual trips arising from a particular infrastructure project.
Consumer surplus	Q24	An important measure of benefit of infrastructure projects is the degree to which they reduce vehicle operating costs
Economics	Q50	Limiting or constraining car access will have a severe impact on the local economy (local retail/business).

Storyline	Question #	Question
Infrastructure investment brings growth	Q26	Investment in all the main types of transport infrastructure (roads, railways, airports) contributes to growth of employment opportunities.
Infrastructure investment brings growth	Q27a*	Investment in heavy fixed rail infrastructure (trains) Investment in light rail infrastructure (trams etc.) Investment in road infrastructure Investment in bus infrastructure Investment in infrastructure for walking Investment in infrastructure for cycling Investment in transport interchange facilities
Infrastructure investment brings growth	Q28	Providing travel time savings for firms through new road infrastructure projects is an efficient means of government economic assistance to business.
Freight - Just-in-time logistics	Q29	Today's logistical methods in industry, based on the principle of 'just-in-time' delivery, mean that more road investment is needed to meet increased demand for freight delivery travel.
Separate markets	Q30	The design of [city] transport system must recognize that separate markets exist for different modes of travel.
Separate markets	Q31	Different travel markets are best served by different specialist government transport agencies (e.g. roads, railway or tramway instrumentalities) or private companies (e.g. providing train/tram services or managing toll road projects).
		Town Planner's Tale
Travel time reduction	Q33	Town planning should seek to reduce the time travelled between the home and other destinations within the urban system to which people choose to travel.
Low density can only be served by roads	Q34	There has to be a strong emphasis on roads and provision for the private car in Australia's cities because of their low gross density.
Australian Conditions – low density = roads	Q52	A transport system suitable for the specific conditions of [city] cannot be achieved without continuous large scale investment in improving the road infrastructure.
Locational equity	Q35	Investment in roads in outer areas of [city] is necessary to compensate for the lack of public transport opportunities in these areas (good public transport service being uneconomical to provide in such areas).

* Question for Q27 set: We would now like to understand your opinion about the contributions different forms of transport infrastructure investments make to growth in employment opportunities (both directly and indirectly). On a scale of 0 to 20 please rate the following in terms of their contribution to growth in employment opportunities: [0 is zero contribution, 20 is the maximum contribution public investment could reasonably make]

Balance between mobility and amenity	Q36	In order to provide a high level of mobility for the majority, the amenity of some areas of [city] may have to be accorded a secondary priority.
Cross-metro travel	Q37	Alternatives to the freeway are needed in [city] to provide better opportunities for cross-metropolitan travel.
Inner freeways	Q10	Building new [motorways/ primary distributor roads] in inner urban areas involves social and environmental costs that are too heavy for communities to bear.
Outer freeways	Q11	Building [motorways/ primary distributor roads] in outer urban areas where land reservations have been established is acceptable because it involves minimal social and environmental costs.
CBD vitality	Q38	The continuing vitality of [city] Central Business District depends on good road connectivity with the suburbs – bringing in customers to the CBD retailers and workers to CBD offices.
Urban Corridors	Q39	Suburban development should be concentrated in spatial corridors around transport spines.
Growth poles	Q40	In the outer suburbs better roads are needed to link households with locations of concentrated employment.
Integrated transport & LU	Q8	In planning a city, transport infrastructure should provide an integrated land use and transport system.
Transport Planning	Q41	The location of transport infrastructure (new rail lines, new roads etc) investment should determine the location of development.
Transport Planning	Q42	The need to ensure all new development is located where it is accessible by all modes of transport should take priority over other competing planning considerations (in the context of an existing city).
		Sustainability Tale
Ecological sustainability	Q55	An ecologically sustainable [city] cannot be achieved while governments continue to favour investment in road transport over public transport and other modes such as walking and cycling.
Resource Efficiency	Q56	Because fossil fuel is becoming scarce and more expensive, cities need to adapt their transport systems to make the best of alternatives to private car transport.
Resource Efficiency	Q57	The way that car use and road space is paid for by the user compared with the way in which public transport is paid for (e.g. in fares collected at each journey) serves as an incentive to use roads.
Roads as 'limited public goods'	Q58	Freedom of movement for all people and freight in cities cannot be achieved in the short to medium term without severe increases in traffic congestion and loss of lifestyle and environmental values in the longer term.
Time saving	Q59	Economic benefits can be achieved by saving time by reducing unnecessary travel.

Storyline	Question #	Question
Time saving	Q60	Economic benefits to firms can be achieved by reducing unnecessary freight transport through advanced logistical planning
Time saving	Q61	Time can be saved for many people by working at home or shopping from home ('telework and e-commerce').
Health Costs	Q62	Because the current costs to life and public health from road transport are so high, a change of priorities for investment is necessary.
Travel as cost	Q68	The full costs of travel (in terms of lives lost and injuries from road accidents, health damaged from pollution as well as the isolation of those left without car access in a car-dependent city) should always be accounted for in any benefit cost analysis.
Travel as cost	Q69	The full costs of car use including congestion, costs to life and health, pollution, and greenhouse emissions, should be paid by the car user.
Roads deficit	Q66	There is a 'roads deficit' (the costs to government arising from road use are at present significantly greater than the receipts to government in taxes and charges for road use).
Car dependency	Q63	[city] has become increasingly 'car-dependent', meaning that the only viable transport mode choice for most people is use of the private car.
Demand Management	Q64	Reducing the need to travel (on all modes) must be a key part of any transport or urban planning strategy
Demand Management	Q65	Managing demand for car travel – and seeking to reduce demand – should be a necessary part of any urban transport strategy today.
Transport Planning	Q67	The public should be allowed to choose collectively what sort of city they desire, and then a transport system should be planned that will create the kind of city desired.
Symbolic values	Q70	People choose cars over other modes of transport because of the symbolic values they embody (e.g. sex appeal, escape, exciting lifestyle, wealth, freedom) at least as much as for the convenience they provide.
Symbolic values	Q71	Changing travel behaviour through economic incentives and other means will lead to changed valuations of modes of transport [following the theory of cognitive dissonance in which it is argued that people generally tend to reduce inconsistency between their values and their behaviour. Thus changing behaviour leads to a change of values].

		Other Minor Storylines
Fixed rail obsolete	Q9	Fixed rail transport systems cannot provide the flexibility, privacy and choice necessary to meet much of today's travel demand.
Community pressure	Q46	Most people in [city] want more and better roads [the majority].
Community pressure	Q47	Most people are satisfied with the quality of public transport service in [city] (the majority).
Community pressure	Q48	Most people are satisfied with the level of cycle infrastructure in [city] (the majority).
Community pressure	Q49	Most people are satisfied with the level of pedestrian infrastructure (extent and condition) in [city].
Community pressure	Q53	People who oppose new road infrastructure projects are often NIMBY activists seeking to protect their local interest in land and are not representative of the wider community.

Index

Page references in **bold** type refer to tables; those in *italic* type refer to figures.